A Guide to Materials Characterization and Chemical Analysis

A Guide to Materials Characterization and Chemical Analysis

Edited by
John P. Sibilia

VCH

Dr. John P. Sibilia
Analytical Sciences Laboratory
Allied Corporation
P.O. Box 1021R
Morristown, NJ 07960-1021

Library of Congress Cataloging-in-Publication Data

A guide to materials characterization and chemical
    analysis.

   Bibliography: p.
   Includes index.
   1. Chemistry, Analytic.   2. Materials—Testing.
I. Sibilia, John P.
QD75.2.G83   1988   543   88-14287
ISBN 0-89573-269-6

British Library of Congress Cataloging-in-Publication Data

A guide to materials characterization and
    chemical analysis.
    1. Materials. Characterization & chemical
    analysis
    I. Sibilia, John P.
    620.1'12

    ISBN 0-89573-269-6

Printed in the United States of America.

ISBN 0-89573-269-6 VCH Publishers
ISBN 3-527-26867-7 VCH Verlagsgesellschaft

Distributed in North America by:

VCH Publishers, Inc.
220 East 23rd Street
Suite 909
New York, New York 10010

Distributed Worldwide by:

VCH Verlagsgesellschaft mbH
P.O. Box 1260/1280
D-6940 Weinheim
Federal Republic of Germany

Preface

The progress in the development of analytical instrumental techniques has been dramatic over the last twenty years. With the coming of age of computer hardware and accompanying software, advances in solid state electronic circuitry and the development of more sensitive detection systems, materials characterization techniques have given impetus to technological advances in optical-electronics, chemicals development, polymers, the biosciences and materials development in general.

This book was written to provide a guide to anyone interested in the characterization of materials with emphasis on analysis of chemicals, polymers, ceramics, metals and composites. Its goal is to provide the novice or student with the salient features of modern materials characterization and analysis techniques. It is also aimed at providing a review for the experienced investigator and expanding the scope of knowledge of those experienced in only a limited number of characterization techniques.

Since 1969 the Corporate Technology laboratories of Allied-Signal Inc. have provided their various research, development and production facilities with a state-of-the-art capabilities guide in the analytical sciences. Since that time four editions of an "Analytical Sciences Skills" booklet have been published.

The most recent Allied-Signal booklet was expanded and broadened in scope resulting in the present work, "A Guide to Materials Characterization and Chemical Analysis."

The introductory chapter gives a brief expose on how one might proceed in utilizing the techniques described in subsequent chapters. Each chapter in turn describes the use, sample requirements, principle, some typical applications, limitations and some general references of the respective techniques.

The authors are all skilled in their particular disciplines and are all current or recent members of the Corporate Analytical Sciences Laboratory which is located in Morristown, NJ, as part of the Corporate Technology organization of Allied-Signal Inc.

There have been contributions through the years to much of the current text by many other former Allied-Signal employees. Although it would be difficult to acknowledge all past contributors, there are several whose efforts were significant in bringing the current book to its final state. Most notable was the effort of Dr. Arthur R. Paterson, the former Director of the Allied Corporate Analytical Sciences Laboratory and since retired. Dr. Paterson's initiative and guidance led to the first edition of Allied's "Analytical Sciences Skills Booklet."

Since that time the late Dr. Willis J. Potts played a key role in the publication of the 3rd edition of the Allied Skills Booklet. Dr. Edith Turi has also played a key role, especially in contributing her editorial ideas to the 4th edition of the Allied-Signal booklet, as well as being an author of the Thermal Analysis chapter in the current work. Also deserving special recognition is Mrs. Anne-marie C. Reimschuessel who in addition to coauthoring the chapter on micros-copy was able, through her dedicated and untiring efforts, to help edit and finalize the 4th edition of the Allied-Signal booklet.

Finally, I would like to acknowledge the support of Dr. L. James Colby, Senior Vice President of Corporate Technology, and Dr. Lance A. Davis, Vice President of Research and Development, Corporate Technology, at Allied-Signal Inc., in bringing the current work to completion.

John P. Sibilia

Contents

Chapter 1

An Introduction to Materials Characterization and Chemical Analysis

John P. Sibilia

Organic Structure Analysis 2
Inorganic Structure Analysis .3
Trace Analysis 3
Polymer Characterization 4
Modeling and Scientific Computation 6
Problem Solving 7

Chapter 2

Molecular Spectroscopy

John P. Sibilia, Willis B. Hammond, John S. Szobota

Infrared Spectroscopy 13
Raman Spectroscopy 19
Ultraviolet, Visible, Near Infrared Spectroscopy 24
Nuclear Magnetic Resonance in Solutions 27
Nuclear Magnetic Resonance in Solids 33
2D-Nuclear Magnetic Resonance Techniques 36
Electron Paramagnetic Resonance Spectroscopy 40

Chapter 3

Mass Spectrometry

R. Donald Sedgwick, David M. Hindenlang

Low and High Resolution Mass Spectrometry 45
Chromatography-Mass Spectrometry 50
Fast Atom Bombardment 53
Tandem Mass Spectrometry 57

Chapter 4

Chromatography

James M. Hanrahan, Mina K. Gabriel, Richard J. Williams,
Milton E. McDonnell

Gas Chromatography 61
Liquid Chromatography 66
Supercritical Fluid Chromatography 71
Ion Chromatography 76
Gel Permeation Chromatography 81

Chapter 5

Chemical Analysis, Electrochemistry and Atomic Spectroscopy

Daniel E. Bause, Richard J. Williams, Kurt Theurer

Classical Chemical Analysis 85
Functional Group Analysis 87
Combustion Analysis 89
Ion-Selective Electrode Analysis 91
Radioactive Tracer Analysis 94
Polarography and Voltammetry 96
Isotachophoresis Analysis 99
Atomic Absorption Spectroscopy 101
Emission Spectrographic Analysis 104
Flame Emission Spectrometry 107
Inductively Coupled Argon Plasma Emission Spectroscopy 109
Neutron Activation Analysis 112

Chapter 6

X-Ray Analysis

Franz Reidinger, N. Sanjeeva Murthy, Steven T. Correale

X-Ray Powder Diffraction 115
X-Ray Diffraction-Polymers 119
Small-Angle X-Ray and Neutron Scattering 124
X-Ray Absorption Spectroscopy 127
X-Ray Fluorescence Spectroscopy 131
Synchrotron X-Ray Sources 133

Chapter 7

Microscopy

Annemarie C. Reimschuessel, John E. Macur, Jordi Marti

Optical Microscopy 137
Scanning Electron Microscopy/Electron Probe Microanalysis 142

Scanning/Conventional Transmission Electron Microscopy 150
Automatic Image Analysis 160
Microscopy Specimen Preparation Techniques 163

Chapter 8

Surface Analysis

Anthony J. Signorelli, Edgar A. Leone, Roland L. Chin

X-Ray Photoelectron Spectroscopy 167
Scanning Auger Microscopy 177
Secondary Ion Mass Spectrometry 185
Ultraviolet and Bremsstrahlung Isochromat Spectroscopy 193
Angular Dependent X-Ray Photoelectron Spectroscopy 197
Scanning Tunneling Microscopy 200

Chapter 9

Thermal Analysis

Edith A. Turi, Yash P. Khanna, Thomas J. Taylor

Thermogravimetric Analysis 205
Differential Thermal Analysis and
    Differential Scanning Calorimetry 211
Thermomechanical Analysis and Dilatometry 217
Thermal Conductivity 222
Dynamic Mechanical Analysis and Sonic Modulus 223
Dielectric Thermal Analysis 227

Chapter 10

The Viscosity and Molecular Weight of Polymers

Abraham M. Kotliar, Milton E. McDonnell, Eugene K. Walsh

Rheology of Fluids 229
Mechanical Spectrometry 233
Molecular Weight of Polymers 235
Colligative Properties of Polymers 235
Viscosity of Polymer Solutions 238
Classical Light Scattering 241
Crosslink Density 243
Field Flow Fractionation 245

Chapter 11

Physical Properties of Particles and Polymers

Milton E. McDonnell, Eugene K. Walsh

General Methods for Particle Size Analysis 251
Photon Correlation Spectroscopy 255

Gas Adsorption 258
Mercury Intrusion Porosimetry 261
Cohesive Energy Density 264
Surface Energy of Solids 266
Dilatometry 269

Chapter 12

Physical Testing

Igor Palley, Anthony J. Signorelli

Mechanical Properties of Materials 273
Fatigue Testing 278
Instrumented Impact Analysis 282
Fracture Toughness of Materials 285
Gas and Liquid Permeability 288

Chapter 13

Scientific Computation

*Sheldon Eichenbaum, Kamal Sarkar, Daniel H. Stevans, David Y. Hsieh,
Gregory J. Czerwienski, Steven E. Sund, Eli Rosenthal,
Willis B. Hammond*

Applied Finite Element Analysis 291
Computional Fluid Dynamics Modeling 294
Dynamic and Steady State Material Process Modeling 297
Molecular Modeling 299

Conversion Factors **305**

Index **307**

CHAPTER 1

An Introduction to Materials Characterization and Chemical Analysis

John P. Sibilia

The Guide to Materials Characterization and Chemical Analysis describes the capabilities of a modern, well-equipped analytical sciences laboratory.

Approximately seventy-five techniques and general methodologies are included in the guide. The use, sample requirements, principle, applications and limitations of each technique are described. In addition, some general references on each technique are given for further reading. Each technique is explained sufficiently to give one with a general chemical/physics/materials background an understanding of how such a technique may be utilized. This kind of information could be useful as a starting point for one who would like to conduct research investigations in depth in any particular technique. However, for those interested or involved in problem solving of a general nature, it is important to know how to utilize a variety of the techniques. In fact most modern analytical sciences laboratories will use a number of techniques in order to conduct materials characterization/problem solving investigations in the most accurate, efficient, time saving manner possible.

In the following paragraphs are given a number of "typical" problems which demonstrate the analytical approaches used in order to most efficiently use the capabilities described in this book. Although the number of examples one could give is endless, the limited number described should provide at least a flavor for problem solving analyses. It is hoped that as one reads these typical examples the chapters which describe each technique in more detail will be referred to frequently.

ORGANIC STRUCTURE ANALYSIS

When an organic chemist carries out a chemical reaction to synthesize a pharmaceutical chemical, a monomer, a herbicide or an intermediate chemical for any end use, he is concerned with at least two major questions. Did he synthesize the material of choice and what is the purity?

The first step in the analysis might be to obtain the infrared (IR) spectrum of the compound. This would provide information on the functional groups in the molecule, and if the molecule has previously been prepared, the IR spectrum can serve to make a positive identification of the compound. Some estimate of purity can also be obtained from the spectrum. If the compound is sufficiently soluble (of the order of 1%) then high resolution nuclear magnetic resonance (NMR) spectrometry can also provide information about the functional groups in the molecule as well as the placement of structural moieties in relation to each other. Again information about the purity of the material can be obtained.

Raman spectroscopy can be used to supplement the IR data and would be especially useful if the material to be analyzed was in an aqueous solution.

If the compound of interest was sufficiently volatile and thermally stable, gas chromatography (GC) could give a good quantitative estimate of the relative purity of the sample. If the compound of interest was positively identified by gas chromatographic techniques, then a rigorous quantitative analysis could readily be accomplished by gas chromatography with the use of standards.

High performance liquid chromatography (HPLC) can be applied if the sample is not very volatile or unstable under GC conditions. Again quantative information on the purity of the sample can be obtained by HPLC techniques.

In many cases the peaks observed in GC or HPLC analyses are not readily identifiable. The next step would be to "trap" out the peaks of interest in sufficient quantity to obtain IR and NMR analyses. However, many techniques are now coupled in one instrument such as GC-MS (mass spectrometry) or HPLC-MS or GC-IR.

The GC-MS or HPLC-MS instruments allow for separation *and* identification of the unknown peaks by accurate analysis of the mass spectrum of the compound. In many cases MS analyses can be conducted directly on the reaction mixture without any previous separation.

Once a compound has been purified and identified by spectroscopic techniques a final elemental analysis may be obtained by chemical methods as further confirmation of the chemical structure.

The schemes described can be readily applied to liquid, solid and gaseous organic compounds.

INORGANIC STRUCTURE ANALYSIS

Analysis of inorganic compounds would proceed in a manner similar to that described for organic compounds with some notable exceptions. Infrared spectroscopy and especially Raman spectroscopy are particularly useful techniques in identifying inorganic as well as organic structures. Some inorganic compounds are sufficiently volatile to obtain useful information from mass spectral analysis. However, by nature, inorganic compounds are not very volatile and in general not very soluble in organic solvents. Therefore techniques such as GC and HPLC are not readily applicable to inorganic analyses. Ion chromatography (IC) is a technique which is extremely useful for identifying inorganic ions. Both qualitative and quantitative analyses of both anions and cations can be achieved by IC techniques. Classical chemical techniques play a more important role in the identification and analysis of inorganic compounds than in organic analysis. Elemental analysis is extremely important in the identification of unknown chemical structures. Wet chemical, X-ray fluorescence (XRF) and other emission analysis techniques are used for elemental analysis of inorganic compounds. Specific ion electrode analysis is also an important technique in the analysis of a number of inorganic anions and cations.

X-ray diffraction (XRD) analysis is a very important tool not only in compound identification but in identifying crystal structures and in crystalline phase analysis. For example two active areas in inorganic research are advanced ceramics and superconductors. While all the techniques described for inorganic analyses are important in compound identification, XRD is especially important in identifying both the chemical composition and crystalline phase of these important classes of new inorganic compounds.

Thermal analytical techniques such as differential thermal analysis (DTA), differential scanning calorimetry (DSC) and thermal gravimetric analysis (TG) are all used on a routine basis for identifying and analyzing inorganic compounds, as well as organic compounds. These techniques provide information about the melting points, other phase transition temperatures and thermal stability of both inorganic and organic compounds.

In many cases particle size, surface area and porosity analyses provide useful characterization information about inorganic materials.

TRACE ANALYSIS

Trace analyses are important in: demonstrating compliance to FDA (Food and Drug Administration), EPA (Environmental Protection Agency), and OSHA (Occupational Safety and Health Administration) regulations; in the forensic sciences; in identifying contamination of electronic parts; in determining monomer and solvent purity and in assaying small amounts of material that can affect the performance or quality of a product.

Many of the techniques described in the book can be used to analyze at the ppm level, the ppb level and even at the sub ppb level. However in many cases it is necessary to prepare a concentrated solution of the impurity or trace component before an analysis is made. For example before a new food packaging material is allowed to be sold to consumers, the FDA requires that no toxic components in the package be allowed to migrate into the food. In order to verify that the packaging material is safe, it is necessary to conduct a series of extractions in food simulating solvents. The extraction temperatures correspond to the maximum temperatures that the packaging films will see in processing.

The extracts would then be concentrated by removing solvent by evaporation. In this way a concentration of 10 fold or more can be achieved. Such solutions would then be analyzed by GC, HPLC and MS to identify and measure components in the solution to well below the ppm level.

In general, concentration techniques are a good way of increasing the sensitivity of analytical methodology.

Some of the most commonly used trace elemental analysis techniques are AA (Atomic Absorption Spectroscopy), OE (Optical Emission Spectroscopy), ICAP (Inductively Coupled Argon Plasma Spectroscopy), Ion Selective Electrode Analyses, Neutron Activation Analyses or coupled techniques such as ICAP-MS.

Microscopy techniques such as optical, scanning electron, microprobe and transmission electron microscopy are all useful as trace analysis techniques, especially in the case of inorganic compounds. Surface analytical techniques such as x-ray photoelectron, auger and secondary ion mass spectrometry are all useful as both qualitative and quantitative trace elemental techniques.

POLYMER CHARACTERIZATION

There are a number of very different features about polymers that makes their characterization different from that of "small" molecule characterization. A polymer, unlike a pure small molecule, contains molecules of different molecular weight. Therefore a polymer molecular weight represents an average of the distribution of the various molecules with different molecular weights. Some molecules may be branched and the end groups of the polymers may be very different from the repeating unit. A bulk of polymer sample can contain residual monomer as well as other low molecular weight "oligomeric" species. Some polymers are amorphous and cannot crystallize into more ordered structures. These materials can be characterized in terms of a glass transition temperature (T_g). This is the temperature where the onset of large scale molecular motion occurs. Other polymers are "semicrystalline" and are capable of forming three dimensional ordered structures. Such semicrystalline polymers exhibit crystalline melting points as well as glass transition temperatures and exhibit x-ray diffraction patterns characteristic of their structures. There are many other "special" features about polymers which cannot be described in this limited writeup. Some of these are described in further detail in subsequent chapters.

When a newly synthesized polymer requires characterization or if a laboratory wants to characterize a competitive polymeric product, there are a number of steps which should be followed. First the viscosity, molecular weight and perhaps the molecular weight distribution should be obtained on a newly synthesized material to determine if in fact the material is polymeric. Then an infrared spectrum and possibly an NMR spectrum are obtained in order to identify functional groups or make a complete identification in the case of a known polymer. Thermal analytical data can also aid in the identification and characterization of the material. DTA or DSC can be used to obtain a melting point, a T_g and a measure of the thermal stability of the polymer. TG can also yield information on the thermal stability of the polymer as well as determine the presence of volatile components such as solvents, residual monomer or oligomers.

XRD analysis of the polymer will determine if the polymer is partially crystalline. XRD can also be used to help identify the chemical structure of the polymer as well as provide information on the preferred conformation of the polymer chains in the crystalline regions.

Optical microscopy is a supporting technique which also provides information on the crystalline nature of the polymer and its melting point. It is also one of the first techniques to be used to help describe the "morphology" of a polymer. Morphology is the term applied to the structural features which describe the size, shape and perfection of lamellae and crystalline aggregates and the way in which these aggregates form larger structures called spherulites. Electron microscopy and XRD techniques are needed to describe the submicron morphological features of polymers such as lamellae and crystallites while optical microscopy is used to study larger morphological features such as the size, and nature of spherulites in a polymer.

There are many physical properties and tests which can be conducted on polymers. The tests are usually designed to obtain fundamental property information or information about the final end use performance or behavior of the product. Final states of polymers are usually films, fibers, molded parts or coatings.

Important common properties of all these forms are the strength and modulus or stiffness of the material. Such properties can be obtained with a tensile tester which measures the force to stretch a material until it breaks.

Important specific properties of films and molded containers are those relating to the permeability of moisture and various gases. Oxygen and moisture permeability are the two most common permeability tests performed on films and molded bottles. These properties are particularly important for plastics used in the food and drug packaging industries.

Special properties relating to fibers are their surface characteristics. Dyeing, staining and soiling are all properties which are affected by the surface characteristics of fibers. Optical microscopy, scanning electron microscopy (SEM) and transmission electron microscopy (TEM) are all techniques useful for studying the physical characteristics of fiber surfaces. Electron probe microanalysis (EPMA) is an elemental identification technique usually associated with SEM. When electrons impact the surface, X-rays are emitted which are characteristic of the elements near the surface. Elements within the first 10000Å or so can be analyzed in this manner. X-ray photoelectron spectroscopy (XPS) also known as Electron Spectroscopy for Chemical Analysis (ESCA) and secondary ion mass spectrometry (SIMS) can be used to examine the elemental compositions of fiber, film and molded plastic surfaces to depths less than 20Å. These techniques are particularly helpful for studying surface chemistry, adhesion related problems and for looking at very thin surface layers of additives or contaminants.

Important special properties of coatings are their ability to adhere to a surface and their surface smoothness. Microscopy and surface energy techniques are helpful in characterizing these properties.

MODELING AND SCIENTIFIC COMPUTATION

An important part of any materials characterization effort should include computational modeling. Computational modeling can be applied in a variety of ways to a variety of problems. In the last chapter of this book four types of modeling are described. They are finite element analysis, fluid dynamic modeling, dynamic and steady state material process modeling and molecular modeling.

All the methods require input of physical-chemical-property data to help describe the energetics, heat transfer, flow properties, kinetics, molecular interactions, etc. These methods in turn can provide new information on the way in which materials, working parts or processes function.

For example if one wanted to develop a model to predict stresses and strains in a fabricated part when subjected to a load, a series of steps would have to be followed.

The part would first have to be defined in terms of its geometry, the physical properties of the material, the stress-strain behavior of the material and the effects of environmental conditions (termperature, humidity, etc.). Since this can result in a considerable number of equations which would be extremely complex to solve, an approximation is made. The part is divided up mathematically into a series of finite elements which are connected through nodes. It is now easier to solve the problem for a finite number of connected elements than for a continuum. All these factors can be expressed in a mathematical procedure or algorithm which can be processed by a computer.

The final results allow one to determine the stress and strain at any point in the fabricated part when the part is subjected to a load at a given temperature. The accuracy of the results will depend on how accurately the model defines the system, how appropriate is the particular algorithm which was used and how accurate are the physical data used for input.

PROBLEM SOLVING

One of the key functions of a modern analytical sciences or materials characterization laboratory is to solve problems or "troubleshoot". The kinds of problems can be extremely diverse and in fact this is one of the most attractive reasons for scientists to enter these fields. Materials characterization problems can arise in research, development, production and technical service. The entire field of forensic science is one of problem solving. Many of the techniques described in this book would be used in a modern forensic laboratory. Many problems in the electronics field such as corrosion of parts, failure of a connection, contamination of a semiconductor or change in electrical properties in general, can all be traced back to some physical or chemically related problem of the system. Again many of the techniques and methods described in this book are currently being used for problem solving in the electronics field.

The chemical and plastics processing industries rely heavily on analytical and material characterization techniques to develop and maintain efficient processing. Of course quality control laboratories use a number of the techniques and methods described, however, in most cases the simpler, less capital intensive methodologies are used. Heterogeneous catalytic chemical processes are very dependent on chemical-materials problem solving techniques. The particle shape, particle size and size distribution, surface area and pore size distribution of a catalyst are all important characteristics affecting catalyst performance. If the catalyst is multicomponent, for example a platinum on carbon reduction catalyst, then the content of platinum and the way it is distributed on the carbon surface will affect its performance. A common problem in heterogeneous catalytic processing is poisoning of the catalyst. Identification of the poison and determining how the poisoning occurred are common trouble-shooting problems.

All the environmental problems associated with chemical processing can fall into the category of troubleshooting. Odors in plant air, foreign odors from a chemical or plastic part, decomposition products from an extruded part, plant air effluents, and waste water effluents are all areas requiring the attention of chemical-materials characterization techniques.

Problems in the pharmaceutical area might involve a drug formulation losing potency on aging or a problem with the packaging. In the medical diagnostics field some formulations may lose their sensitivity or give false results, again problems which lend themselves to troubleshooting capabilities.

The number and kinds of problems are endless. In order to give a feeling for how one might approach such troubleshooting problems, three examples are given. The first is a plastic film problem, the second is a thermoset polymer problem and the third is a problem dealing with the failure of an electronic component.

Let us consider first a complex problem which would involve both troubleshooting and competitive product analysis. Consider the case where two companies are producing a similar coextruded film used in food packaging. A technical service representative describes both products as a nylon/ionomer/polyethylene multilayered film. However in one film the adhesive bond between the nylon and ionomer has failed while in the competitive product the multilayered film functions well. Upon receiving film samples of both product A (the failed material) and product B (the good competitive material) the analyst or investigating scientist would first confirm the observations of the technical service representative. Visual and optical microscopic examination of cross sections of the films would first confirm that in fact delamination does occur in A between the nylon and ionomer but does not occur in B. Further, polarized light optical microscopy would determine the number of layers, the thickness of the layers and a tentative identification of the layers through hot stage melting points. Additionally, information about the spherulitic/crystalline layers and the nature of the interface between the layers can be obtained by the optical microscopy techniques.

DTA analyses would then be performed to yield melting points and possibly glass transition temperatures of the polymers.

If the nylon had a melting point of about 220°C it would be tentatively identified as nylon 6. If the melting point was about 260°C it could be nylon 66. Melting points of about 100°C and 130°C respectively would correspond to the ionomer and polyethylene layers.

At this point infrared spectroscopy would be used to further confirm the identifications. An infrared transmission analysis where the infrared radiation passes through all the layers of the composite film would help to establish that three components, nylon, ionomer and polyethylene were actually present. However, specific identifications as to type of nylon, ionomer and polyethylene would still be a problem since the spectra of the three compounds would be superimposed making the spectral features of the individual components difficult to discern. Two other approaches can then be used to obtain spectra of the individual components. The older, more classical approach would require chemical separation of the layers, since coextruded or laminated films are normally difficult to separate mechanically.

Trifluoroethanol or formic acid are good solvents for nylon but will not dissolve the polyolefins (ionomer or polyethylene). Therefore a multilayer film with nylon as an outer layer can be stripped of the nylon by immersing the film in trifluoroethanol at room temperature. After the nylon is dissolved, a thin film of the nylon solution is cast on a glass plate and the solvent evaporated at room temperature or slightly above. The dried nylon film can normally be removed from the glass plate as a single coherent sheet. An infrared transmission spectrum of the cast film will now allow a positive identification of the nylon type to be made since no interfering peaks from the polyolefins are present in the spectrum. A similar extraction can be performed on the original film or remaining composite film by using toluene to dissolve the polyethylene. A casting of the toluene solution, drying and subsequent IR analysis will allow for positive identification of the polyethylene as well as provide information about the type of polyethylene (degree of unsaturation, branching and possible molecular weight). Molecular weight, molecular weight distribution and branching can also be obtained by techniques such as osmometry, light scattering and gel permeation chromatography once the material has been solubilized. The ionomer film, the only component remaining after dissolution of the nylon and polyethylene can then be analyzed directly by the IR for identification as an ionomer as well as the degree of neutralization in the acid component of the ionomer. A more direct and faster way of obtaining similar and actually additional information is by using an infrared microscope. In this procedure a thin ($\sim$ 25μm) cross section of the composite film is mounted in the stage of the infrared microscope such that the IR radiation can pass through each layer without interference from adjacent layers. In this way positive identification can be made by obtaining the spectrum of each layer independently. An additional advantage of this method is that morphological information (crystallinity and orientation) as well as information about the interfaces can be obtained directly on the sample without alteration. Although the dissolution technique previously described has an advantage in allowing solution properties to be obtained, a disadvantage is that it destroys the original morphology and orientation so that the cast film will most likely have a different morphology from the original melt extruded film. Sometimes it is still not possible to make a positive identification of the nylon type even with the methods described. There are a number of commercial copolymers of nylon whose melting points may be only slightly different from those of the pure materials and whose infrared spectra are very similar to the pure nylons. Therefore to finally confirm the nylon type an NMR spectrum on the solubilized nylon would also be helpful. The final definitive method requires a hydrolysis of the nylon with HCl followed by derivatization of the hydrolyzed components in order to make them sufficiently volatile for a GC analysis. A compound such as the dimethylacetal of N,N-dimethylformamide can be used as an effective derivatizing agent for hydrolyzed nylons. A gas chromatographic analysis of the derivatized hydrolyzates will not only positively identify the monomers but provide a composition analysis of the copolymers as well as determine the presence of volatile impurities. HPLC analysis might also be used for analysis of the hydrolyzate products without derivatization.

An ESCA analysis of the delaminated surfaces may yield information about impurities which could prevent adhesion of the layers. Surface impurities detected by this sensitive surface technique might not be detected by any of the techniques described previously.

The number of these analyses to be performed depends upon the amount of information obtained in each analysis, the time required to provide useful information, the costs and the quality of the original information provided.

When all these data are compared, one can conclude how A and B are different and what are the probable causes of the delamination. Some of the possible reasons for the delaminations of A may have been due to differences in composition, thickness, morphology, molecular weight, impurities, etc.

A second type of troubleshooting problem might include strictly organic polymer structural analysis. For example, certain composite fabrications are comprised of phenolic resins as a matrix material. The composites are prepared from a viscous "prepolymer" mixed with various inorganic fillers and finally cured or crosslinked into a molded product. A typical problem associated with this chemistry is that it may not be possible to process or mold the part into a good finished product with acceptable properties.

The first step in determining the causes for poor performance in such a material is to obtain an IR spectrum of the precured composite to determine if any impurities are present. If the filler content is too high, the spectrum of the inorganic components would mask the organic components such that the spectrum would not be very informative. In this case the organic portion should be separated by solvent extraction followed by filtration of the nonsoluble fillers. The soluble portion can then be plated out on a salt window, the solvent removed and the IR spectrum obtained. Again the spectrum should first be examined for impurities. This analysis is facilitated by comparing the spectrum of the "poor" performing phenolic polymer with a comparable spectrum of a "good" performing polymer or at least spectra of "standard" polymers. In addition to impurities as possible causes of the improper curing, concentration of catalysts and functional group concentrations are also important. One of the most important characteristics of the precured polymer is the nature and molecular weight of the oligomeric species. Identification of these features requires GC and HPLC analyses of the soluble portion of the phenolic composite, followed by high resolution MS analyses. Again, when these results are compared with those from a standard or good performing material the differences in catalyst concentration, impurities and oligomer molecular weight will usually account for the differences in performance of the polymers.

A third example of problem solving might involve the malfunctioning of an integrated circuit or electronic part. Corrosion of components, surface impurities and bulk impurities in many cases result in defects which cause the malfunction of the device. Usually these defects are microscopic in nature and require analysis by electron as well as optical microscopy. SEM and electron probe microanalysis as well as optical microscopy are important screening tools in these types of analyses. Additional information may be obtained by scanning auger microscopy (SAM), XPS and SIMS. These techniques will allow for physical examination of parts for defects, i.e., cracks, pits, surface defects. Elemental analysis of the corrosion products or impurities can also be achieved by these techniques.

One specific example of an electronics/material problem involves the failure of a thick film resistor used in a sensor electronic circuit. A thick film resistor consisting of a thin rhodium layer on an alumina substrate was found to undergo a negative resistance drift on aging.

The resistance of such devices is obtained by "laser" trimming the metal until the exact desired resistance is obtained. Initially a SEM study was undertaken to examine the physical characteristics of the area known as the "kerf", which the laser had trimmed. High resolution SEM showed that the path that the laser had cut was not very uniform in depth and did not go completely down to the substrate. Further examination showed small pieces of debris in the kerf. Microprobe analysis of the debris showed that they contained some rhodium particles. Therefore based on the SEM and microprobe analyses it was concluded that the negative resistance drift was due to residual conducting particles in the kerf. Based on these results and microscopic analysis of cross sections, it was further concluded that the laser was not cutting deeply enough and effectively enough to make a clean deep kerf. Improvement in the laser cutting process eliminated the problem.

The problems cited are presented to give one a feeling for how a modern materials characterization laboratory functions. The approach one takes to problem solving is extremely important in determining whether or not a problem will be solved and how quickly the solution can be accomplished. In the following chapters are given the description of a wide variety of analytical and materials characterization techniques. The way in which these techniques are utilized in problem solving are limited only by the imagination of the investigator. It is hoped that the manner in which the material is presented will help serve as an effective teaching guide and will stimulate the interest of those involved in the exciting field of materials characterization and chemical analysis.

CHAPTER 2

Molecular Spectroscopy

John P. Sibilia, Willis B. Hammond, John S. Szobota

INFRARED SPECTROSCOPY

Use

Infrared spectroscopy can be used to identify materials, determine the composition of mixtures, monitor the course and extent of reactions, and provide information useful in deducing molecular structure.

Sample

Materials in the solid, liquid or gaseous state may be studied by infrared spectroscopy. A convenient sample size is several milligrams, but spectra can be obtained from as little as 50 picograms with special techniques.

Principle

Analysis by infrared spectroscopy is based on the fact that molecules have specific frequencies of internal vibrations. These frequencies occur in the infrared region of the electromagnetic spectrum: $\sim 4000 cm^{-1}$ to $\sim 200 cm^{-1}$.

When a sample is placed in a beam of infrared radiation, the sample will absorb radiation at frequencies corresponding to molecular vibrational frequencies, but will transmit all other frequencies. The frequencies of radiation absorbed are measured by an infrared spectrometer, and the resulting plot of absorbed energy vs. frequency is called the infrared spectrum of the material.

Identification of a substance is possible because different materials have different vibrations and yield different infrared spectra. Furthermore, from the frequencies of the absorptions it is possible to determine whether various chemical groups are present or absent in a chemical structure. In addition to the characteristic nature of the absorptions, the magnitude of the absorption due to a given species is related to the concentration of that species.

Applications

Infrared spectroscopy is one of the major tools for obtaining information regarding the structure of molecules. Besides the standard applications of identification and quantitative analysis, there are a number of special techniques and applications. Some of these are listed as follows:

Polymer Chemistry

1. Determination of number average molecular weight through end group measurements in certain polymers.
2. Determination of crystalline index and branching.
3. Estimation of degrees of stereoregularity, conformational analysis and sequence distribution in copolymers.
4. Analysis of degradation products to establish mechanisms of degradation.
5. Orientation studies through the application of polarized radiation.
6. Determination of T_g and phase changes through temperature experiments between $-50°C$ and $400°C$.
7. Study of the surface of a material by attenuated total reflection (ATR) and variation of composition with depth by variable angle ATR.

Organic Chemistry

1. Reaction kinetic studies.
2. Studies of reactive intermediates between $-50°C$ and $400°C$.
3. Conformational and configurational studies on both cyclic and acyclic systems.
4. H-bonding, dipolar attractions and solute-solvent interaction studies.

Inorganic Chemistry

1. The nature of inorganic lattices at different temperatures.
2. Nature of hydration.
3. Formation and reaction of $-OH$ groups in various inorganic glasses.
4. Symmetry properties of the arrangement of atoms through a consideration of group theory.

Fourier Transform Infrared Spectroscopy

For many decades, the work-horse of the infrared laboratory has been the light-dispersive spectrometer. While still in wide use for routine analysis, the dispersive nature of these spectrometers, however, severely limits their application in solving "difficult problems"; e.g., microsamples, strong IR absorbers, and analysis of impurities in most samples at levels below $\sim$0.1 to 1% without special pre-concentration procedures.

During the past decade, Fourier-transform infrared spectroscopy (FT-IR), coupled with the rapid development during this period of a wide variety of sampling accessories designed to make optimal use of the advantages of FT-IR, has come to the forefront as the technique capable of handling both the "routine" and "difficult" problems. A schematic of an FT-IR spectrometer is given in Figure 1.

Fourier transform infrared spectroscopy employs an interferometer in place of a monochromator. This device generates the Fourier transform of the infrared spectrum which is converted to the spectrum itself by a dedicated computer. This approach has the advantages of providing much higher source radiation throughput, increased signal/noise ratio, and higher wavenumber accuracy than is possible with a conventional light-dispersive spectrometer. These advantages can be applied in several ways.

1. Useful spectral information can be obtained from a sample of a microgram or less. Utilizing special techniques, as little as 50 picograms may be analyzed.
2. A spectrum can be obtained in a much shorter time than is possible with a dispersive spectrometer. Thus, spectra can be obtained of transient species or phenomena down to as short a time as $\frac{1}{30}$th of a second.

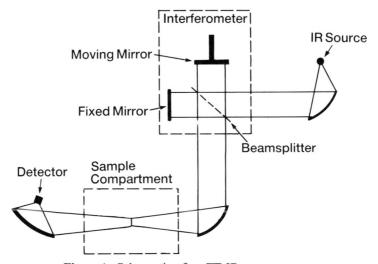

Figure 1. Schematic of an FT-IR spectrometer.

3. A combination of both those advantages makes it possible to obtain infrared spectra of gas chromatographic and high pressure liquid chromatographic cuts as they emerge from the chromatograph, thus providing considerable structural information about the material giving rise to each peak in the chromatogram (cf. GC-mass spectrometry and HPLC).
4. The spectrum can be obtained at very high resolution, which has certain advantages in studying small molecules in the vapor phase.
5. Utilizing microscope accessories, spectra of individual particles or inclusions of the order of 5μm in size can be obtained.
6. The modulation provided by the interferometer allows the use of Photoacoustic Spectroscopy (PAS), whereby the spectra of material may be obtained "as is" (powders, pellets, chunks, etc.). PAS involves infrared absorption in the sample, followed by conversion of the absorbed energy into heat. The subsequent heat-induced thermal expansion in the sample and adjacent media produces a photoacoustic signal when the incident beam intensity is modulated at a frequency in the acoustic range. Microphonic detection of this signal, processed by the normal detector amplification electronics of an FT-IR spectrometer, yields the spectrum.

Examples

In Figures 2–5 are shown examples of nonroutine analyses that may be conducted utilizing infrared spectroscopy. The examples indicate the utility of IR as a surface analytical tool, a quantitative technique for the determination of extremely small amounts of material, and its applicability to polymer studies.

Figure 2 shows the PAS spectrum of a sample of CF_x, contaminated by oxygen containing species, at two different moving mirror velocities corresponding to depths of penetration of ~ 0.25 and 1μm. Transmission IR of the sample as a KBr pellet showed only the C-F stretch at ~ 1200cm^{-1}. Note

the variation of the C-H stretch and the C-F vs. C-C-F band intensities with depth. This illustrates how infrared provides information regarding the specific chemical groups involved in surface contamination.

Figure 3 is the spectrum of an impurity removed from a printed circuit board by washing the board with acetonitrile and evaporating this wash liquid on a KRS-5 ATR crystal. The spectrum obtained from a sample of the suspected rosin contaminant was identical to that obtained from the wash. Utilizing spectra obtained in the same manner from standard solutions of the rosin indicated that the actual sample analyzed was of the order of 20 nanograms.

In Figures 4 and 5 are shown examples of two types of nonroutine analysis that may be conducted on polymeric materials. The examples indicate how

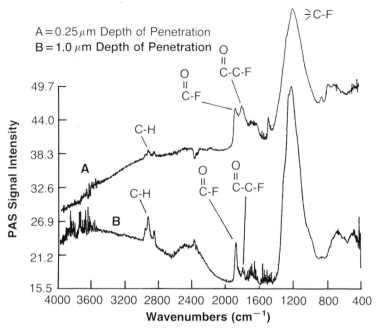

Figure 2. PAS of oxygen contaminated CF_x.

both physical and chemical phenomena can be investigated in the same material. Figure 4 shows how the intensity of the symmetric —CH_2— stretching absorption in a vinylidine fluoride (4%) monochlorotrifluoroethylene (96%) copolymer changes in the region of the glass transition (T_g). This

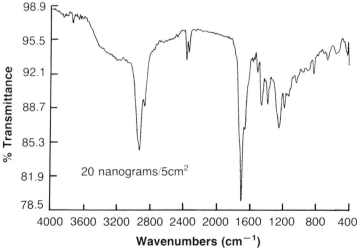

Figure 3. ATR spectrum of rosin contaminant on printed circuit board.

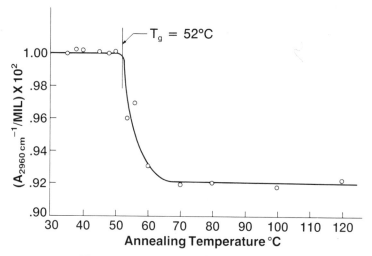

Figure 4. T_g of a halocarbon film by IR.

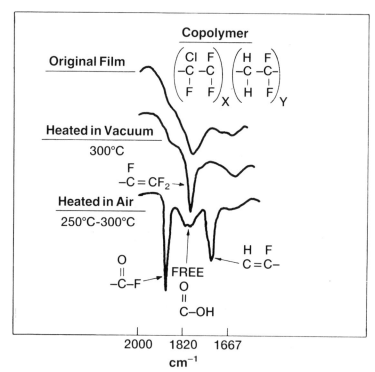

Figure 5. IR spectra of thermally degraded halocarbon film.

illustrates how infrared provides information regarding polymer transitions. In Figure 5 are shown spectra of the same copolymer after various heat treatments. The formation of various functional groups formed in different atmospheres is shown. Results of this type, along with identification of the gaseous products, can give valuable information about the nature of degradative mechanisms in polymers.

Limitations

Routine analyses using solutions, mulls, or specialized sampling accessories, for example, diffuse reflectance or photoacoustic detection, require milligram quantities of material. This requirement has been reduced to the nanogram level with the development of a variety of microanalytical accessories. Infrared detection of chromatographic cuts (both GC and HPLC) has been pushed to the nanogram level. Using matrix isolation techniques, 50 pg of material can now be detected in GC-IR analyses. These detection limits are achievable with commercially available instrumentation.

References

(1) Griffiths, P. R.; de Haseth, J. A., Fourier Transform Infrared Spectrometry; John Wiley and Sons: New York, 1986.
(2) Sibilia, J. P.; Paterson, A. R., J. Polymer Sci., Part C, No. 8, 1965, pp 41–57.
(3) Smith, A. L., Applied Infrared Spectroscopy; John Wiley and Sons: New York, 1979.

Acknowledgment

Figures 4 and 5 are reprinted from reference 2, copyright© 1964 with permission of John Wiley and Sons, Inc.

RAMAN SPECTROSCOPY

Use

Raman Spectroscopy is used to determine molecular structures and compositions of organic and inorganic materials.

Sample

Materials in the solid, liquid or gaseous state can be examined. Normally, tenths of a gram are the minimum sample requirements. However, in some cases microgram quantities can be studied.

Principle

When an intense beam of monochromatic light impinges on a material, scattering can occur in all directions—with the frequency of the scattered light being the same as that of the original light (ν_0). This effect is known as Rayleigh scattering. Another type of scattering that can occur simultaneously with the Rayleigh scattering is known as the Raman effect. It occurs at frequencies both higher and lower than ν_0 and with considerably diminished intensities. The differences, $\Delta\nu$, between the incident and scattered frequencies are equal to the actual vibrational frequencies of the material. Therefore, Raman spectroscopy, like infrared, provides characteristic frequencies of various functional groups. However, since the selection rules governing the allowable transitions are different, some frequencies may be observed in the Raman spectrum which do not appear in the infrared, and vice versa.

A schematic of a Raman spectrometer is shown in Figure 1.

Applications

1. Identification and molecular structure determination of organic and inorganic compounds.
2. Examination of aqueous solutions of inorganic compounds. While H_2O gives rise to intense absorptions in the infrared, making unavailable major regions of the spectrum for identification purposes, it is a poor Raman scatterer, thereby allowing the observation of vibrational transitions in the regions obscured in the infrared.
3. Structural identifications of water-soluble organic compounds such as amino acids.
4. Determination of the nature of lattice vibrations in crystals.
5. Examination of the low vibrational frequencies of materials.
6. Detection of weak infrared frequencies, such as the stretching vibrations of the following groups:

    $-C{\equiv}C-$, $C{=}C$, $-S-S-$, $-C-S-$, $-N{=}N-$, and $-O-O-$
7. Determination of configurational isomers in both the solid and liquid state.

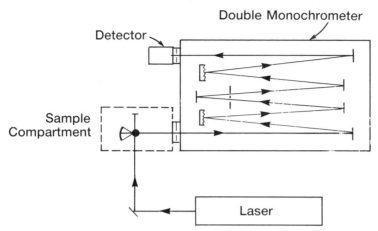

Figure 1. Schematic of a Raman spectrometer.

8. Examination of air and moisture sensitive materials. Materials may be examined in glass containers as prepared or condensed into glass tubes from a vacuum line since the Raman scatter from glass is weak and glass also transmits the laser lines in the visible region. In the infrared, glass absorbs below 2000 cm^{-1} making this region unavailable for analytical purposes by infrared spectroscopy.

9. Use of the rotating Raman cell and/or the low temperature Raman cell to examine highly adsorbing, photosensitive or thermally sensitive materials.

10. Application of the resonance Raman effect to the study of colored or highly conjugated materials. In the case of a material having a distribution of conjugation lengths, the exciting line may be varied across the electronic absorption bands of the species present, thereby selectively resonance enhancing the Raman bands of species of different conjugation lengths.

11. Utilization of microscope accessories allowing the Raman spectra of particles or inclusions of the order of 5 μm in size to be obtained.

In Figure 2 is shown the infrared (top) and Raman spectrum (bottom) of bis-(pentamethyldisilanyl) acetylene which is a simple model monomer for polymeric systems which may exhibit side-chain crystallization without a corresponding crystallization of the main chain. The infrared spectrum provides evidence for the identification of substituent groups on silicon. For example, the CH stretching bands at 2955 and 2896 cm^{-1}, the asymmetric CH$_3$ deformation at 1409 cm^{-1} and the CH$_3$ symmetric deformation at 1246 cm^{-1} are characteristic for CH$_3$ substitution on Si. Furthermore, the mixed mode methyl rock and Si-C stretch region near 800 cm^{-1} shows bands at 838 and 765 cm^{-1} indicative of the presence of three methyl groups bonded

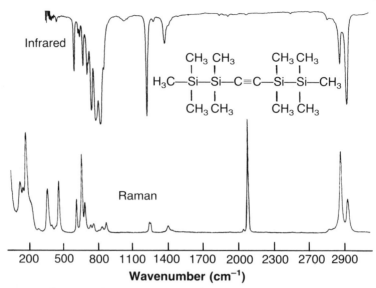

Figure 2. Infrared and Raman spectra of bis(pentamethyldisilanyl) acetylene.

to Si, while the band at 801 cm^{-1} is characteristic of dimethyl substituted Si. The Raman spectrum possesses a very intense band at 2092 cm^{-1} due to $-C\equiv C-$ stretching. This functional group generally gives rise to very weak bands in the infrared and if symmetrically substituted is symmetry forbidden in the infrared. The bands at 464 and 366 cm^{-1} are due to Si-Si stretching while the bands near 200 cm^{-1} and below are due to methyl torsions. Notice that although there are numerous coincidences between the two spectra, the differences in relative band intensities can be dramatic. For example, consider the 2092 cm^{-1} band which is very intense in the Raman spectrum and exceedingly weak in the infrared. Note also the 1052 cm^{-1} band in the infrared, indicative of a small amount of Si-O-Si containing impurity, which is not observable in the Raman.

In Figure 3 is shown the infrared (top) and Raman spectrum (bottom) of Nacconate 80 which is a commercial blend of 80% 2,4- and 20% 2,6-toluene diisocyanate. The infrared spectrum provides evidence for the identification and location of substituent groups on the aromatic ring. For example, the highest intensity infrared band near 2280 cm^{-1} is quite characteristic of the $-N=C=O$ group. The Raman spectrum possesses a high intensity band near 1510 cm^{-1} which is due to a symmetric stretching of the aromatic ring and the $-N=C=O$ groups. Even though there are many coincidences between the two spectra, the differences in relative band intensities can be large. For example, consider the 2280 cm^{-1} band which is very intense in the infrared spectrum and very weak in the Raman effect. This difference in

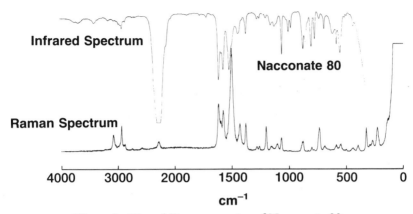

Figure 3. IR and Raman spectra of Nacconate 80.

intensity reflects the different physical origins of the two techniques, and illustrates the complimentary nature of infrared and Raman spectroscopy.

Limitations

The primary limitation of this technique resides in the intrinsic weakness of the Raman effect. Generally, sample requirements are a few tenths of a gram, but with micro-Raman accessories microgram quantities of material can now be studied. In addition, many samples contain minor components or impurities which fluoresce when excited by visible laser radiation. This fluorescence can be sufficiently intense to completely mask the Raman scattered radiation, thereby limiting the range of materials that can be examined. However, recent developments in Raman spectroscopy using near-infrared laser excitation promise to eliminate the fluorescence problem for most materials since most compounds are transparent in the near-infrared region.

References

(1) Hallmark, V.; Zimba, C. G.; Swalen, J. D.; Rabolt, J. F., Spectroscopy, 2, 1987, 40.
(2) Long, D. A., Raman Spectroscopy; McGraw-Hill: New York, 1977.
(3) Szymanski, H. A., Raman Spectroscopy, Theory and Practice; Plenum Press: New York, 1967, 1970, Vol. 1 and 2.

ULTRAVIOLET, VISIBLE, AND NEAR INFRARED SPECTROSCOPY

Use

This broad region of the electromagnetic spectrum finds its principal application in making quantitative measurements of materials in dilute solution. It also has some limited application in the study of molecular structure; the near infrared region can give useful information about OH and NH groups from overtones and combinations of their fundamental stretching frequencies (see Infrared Spectroscopy); the ultraviolet region of the spectrum provides information about conjugated π-electronic systems, especially aromatic systems.

Sample

Liquids and solids are examined in dilute solution where a wide variety of solvent choices are possible. Vapors are examined directly. The sample size required is typically of the order of 1 mg, but in special applications in the ultraviolet region spectral information can be obtained from nanogram amounts.

Principle

The near infrared region (NIR) extends from $\sim$2500 nm to $\sim$750 nm. Light absorptions in this region of the spectrum are inherently weak, and result from excitation of overtones and combinations of fundamental vibrations which are principally hydrogen stretching motions (see Infrared Spectroscopy).

Light absorption in the visible (vis. $\sim$750 nm to $\sim$400 nm) and ultraviolet (UV $\sim$400 nm to $\sim$180 nm) regions of the spectrum result from excitation of electronic states. Generally speaking, the larger the region of the molecule over which π-electron conjugation can occur, the longer the wavelength of the absorption and the more intense the resulting absorption band. Thus, simple alkenes absorb at $\sim$200 nm, while large dye molecules absorb in the visible region.

These three "regions" of the spectrum—NIR, visible, and UV—are conveniently classed together even though the molecular phenomena giving rise to light absorption are different. Instrumentation principles for all three regions are very similar (in fact, two or all three of these regions are often covered by one instrument). Photometry in all three is sensitive and precise. The range of the solvent choices for all is broad (including water for most of the region) and quartz can be used as a window material over the entire range, thus allowing great flexibility for special applications.

Applications

1. Because of its high sensitivity to aromatic systems, photometry in the ultraviolet region is often used as a final step in a wide variety of analytical chemical procedures. The most commonly employed detector in liquid chromatography is ultraviolet absorption at a fixed wavelength.
2. The near infrared region, being highly specific to materials having OH and NH groups, is very often useful for quantitative analysis of such compounds as oximes, lactams, and alcohols. End group analysis (OH and NH), and thus molecular weight values, can be obtained on some polymeric systems.
3. Quantitative analysis of components present at concentrations of ~1% in weakly absorbing samples in powder form has been a very difficult problem to attack by traditional quantitative spectroscopic techniques. Near-Infrared (diffuse) Reflectance Analysis (NIRA) at multiple wavelengths coupled with multivariate statistics has been applied to difficult samples of this type with great success. The convenience of sample handling, computer assistance, and additivity of the near-infrared response make this the ideal technique for weakly absorbing powder samples.
4. Because of the wide variety of solvent choice and the flexibility in making special absorption cells, near infrared, visible, and ultraviolet spectroscopy can conveniently be used to study equilibrium phenomena. As an example, consider the monomer-dimer equilibrium in non-polar solvents for caprolactam:

$$2(CH_2)_5 \begin{matrix} C=O \\ | \\ NH \end{matrix} \rightleftharpoons (CH_2)_5 \begin{matrix} C=O\cdots NH \\ | \quad\quad | \\ NH\cdots O=C \end{matrix} (CH_2)_5$$

It can be shown that the absorbance (A_m) of the monomeric -NH stretching overtone at 1.49 μm is related to the total caprolactam concentration (C) through the following relationship:

$$A_m^2 + \frac{A_m kL}{2K} = \frac{C(kL)^2}{2K}$$

Where K is the monomer-dimer equilibrium constant, k is the absorption coefficient of the 1.49 μm band, and L is the cell path length.

If the tendency to form dimer is low, then K will be small and the equation will reduce to $A_m = CkL$, the normal Beer's Law relationship.

In Figure 1 is shown the non-linear relationship which results when A_m is plotted against the concentration. This non-linearity indicates that the amount of dimer formed at these concentrations is appreciable. In Figure 2 is given the plot of A_m^2 versus concentration. This plot is approximately linear, indicating that K must be large.

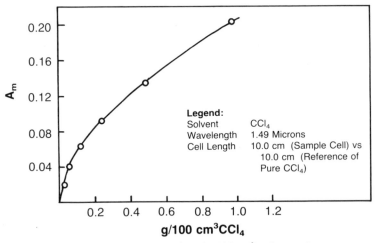

Figure 1. A_m vs. concentration (g/100cm³) of caprolactam.

Limitations

The ultraviolet-visible absorption bands are generally broad and relatively featureless which severely limits the application of this technique to the determination of molecular structures. Near infrared spectra have some utility for identifying O-H, N-H, C-H moieties, but the overtone-combination band spectra are generally too complex for further molecular structure determination. Sample requirements range from a few mg (for standard analytical procedures) down to sub-microgram quantities (for specialized applications such as liquid chromatography detection and microspectrometry).

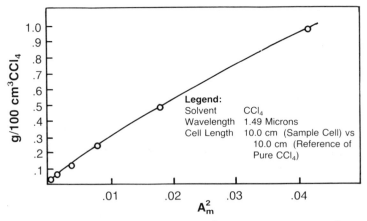

Figure 2. Concentration (g/100 cm³) of caprolactam vs. A_m^2.

Reference

West, W., Techniques of Organic Chemistry (Weissberger, A., ed); Interscience: New York, 1968, vol. IX.

NUCLEAR MAGNETIC RESONANCE IN SOLUTIONS

Use

Nuclear Magnetic Resonance Spectroscopy (NMR) gives detailed structural information on materials containing atoms which possess a magnetic moment, e.g. ^{1}H, ^{13}C, ^{19}F, ^{29}Si, ^{31}P, and many others. Information on the dynamics of molecules in solution can also be obtained.

Samples

High resolution spectra are most conveniently obtained on materials as liquids, or as solutions in appropriate solvents. ^{1}H spectra can be obtained on as little as 10 μg of sample and ^{13}C spectra can be obtained on 1 mg of material in favorable cases. Larger samples will reduce the time needed for data acquisition. Sample requirements vary for other nuclei.

Principle

When a sample containing a nucleus with a non-zero magnetic moment is placed in a strong magnetic field, H_0, the magnetic moment of the nucleus will begin precessing about H_0 with a frequency $\nu = \gamma H_{eff}/2\pi$ where γ is the gyromagnetic ratio, a constant characteristic of the particular nucleus, and H_{eff} is the effective magnetic field experienced by the nucleus. Application of a strong radio-frequency (RF) pulse at a frequency close to ν and perpendicular to H_0 will change the angle of the magnetic moment with respect to H_0 and induce a signal in a detector.

The resulting free induction decay signal (FID) is modulated by the precessing magnetic moment and decays in intensity as the magnetic moment relaxes back to its equilibrium position. Fourier transformation of the time dependent FID converts it to a frequency spectrum containing information on each nucleus. Since H_{eff} is sensitive to the immediate environment of the nucleus in the molecule, nuclei in different environments resonate at different frequencies and give structural information on the molecular environment. The location (frequency) of a particular resonance when compared to the frequency of a reference material is defined as its chemical shift (in parts per million, ppm). The intensity of a resonance is proportional to the number of nuclei in that particular magnetic environment, allowing quantitative measurements.

Interactions between nearby magnetically non-equivalent nuclei result in multiple resonances for each nucleus. This interaction is called spin-spin coupling. The number of resonances and the magnitude of their separation (in Hz) gives important information on molecular structure and spacial relationships within the molecule.

A schematic of a Fourier transform NMR spectrometer is given in Figure 1.

Applications

NMR is one of the most powerful tools available to the chemist for molecular structure determination. It is useful for characterizing organic, inorganic, biological and organometallic molecules including macromolecules such as polymers and proteins. In addition, NMR can measure the rates of chemical reactions and give important information on internal motions in complex molecules. Modern Fourier Transform NMR spectrometers with high field superconducting magnets have dramatically increased the sensitivity and resolution of the technique. In addition to ^1H, ^{19}F, and ^{31}P which are relatively easy to detect by NMR, less isotopically abundant nuclei such as ^{13}C, ^{15}N, ^{17}O, and ^{29}Si, are now routinely studied.

The development of broad band NMR probes that are easily tunable over a range of frequencies now make it convenient to study many more nuclei. Included in this list are: ^2H, $^{6,7}Li$, $^{10,11}B$, ^{23}Na, ^{27}Al, ^{29}Si, ^{33}S, ^{77}Se, ^{117}Sn, ^{195}Pt, and ^{199}Hg.

Some auxiliary features on a modern spectrometer are: (1) Computer controlled variable temperature from -150 to $200°C$ ($300°C$ in certain cases),

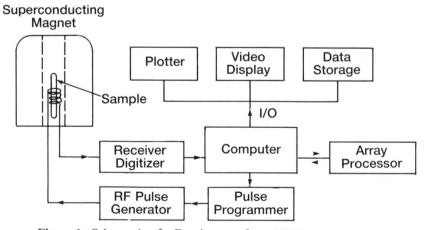

Figure 1. Schematic of a Fourier transform NMR spectrometer.

to aid in the study of temperature dependent phenomena or to increase the solubility of poorly soluble materials. (2)Homo- and heteronuclear spin decoupling, for simplifying complex spectra or increasing sensitivity. (3)A programmable multipulse sequence generator for enhancing spectrometer sensitivity for less common nuclei, for spectral editing, and for generating a variety of 2D NMR pulse sequences. (4)Automated T_1 (spin-lattice relaxation time) and T_2 (spin-spin relaxation time) measurements which can aid in understanding physical properties of polymers.

Various areas of application of NMR follow:

1. Structure elucidation of organic, inorganic and biological compounds.
2. Identification of polymers, copolymers, and biopolymers.
3. Analysis of type and degree of polymer branching.
4. Analysis of copolymer sequence distribution and tacticity.
5. Study of internal polymer motion by T_1 and T_2 measurements.
6. Kinetic studies of reactions at temperatures from $-150°C$ to $300°C$.
7. Identification and analysis of impurities in raw materials and products.
8. Characterization of metabolites and trace materials by selective labeling with ^{15}N, ^{13}C, ^2H, or ^{17}O.

Figure 2 presents the 400 MHz ^1H NMR spectrum of a sugar derivative and illustrates the ability of high field NMR to reduce a complex spectrum to first order. Figure 3 presents the 100 MHz ^{13}C NMR spectrum of high density polyethylene. Low levels of butyl branching as well as terminal

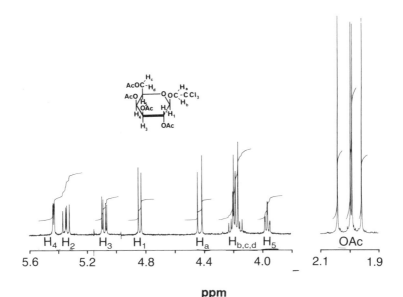

ppm

Figure 2. ^1H, 400 MHz.

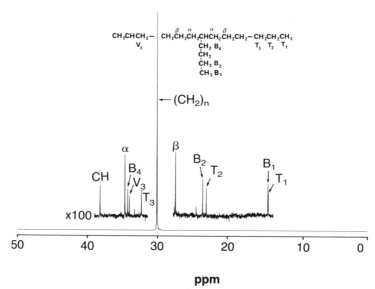

Figure 3. ^{13}C, 100 MHz.

methyl and vinyl groups can be detected and analyzed quantitatively for comonomer content and number average molecular weight. Figure 4 shows the ^{31}P NMR spectrum of ATP, an important component of living systems. ADP, AMP and inorganic phosphate are also detected. Figure 5 illustrates

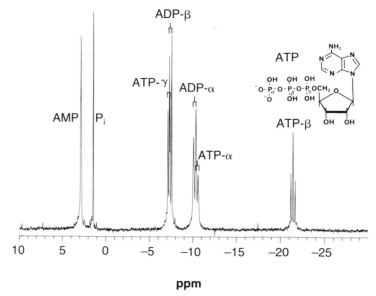

Figure 4. ^{31}P, 162 MHz.

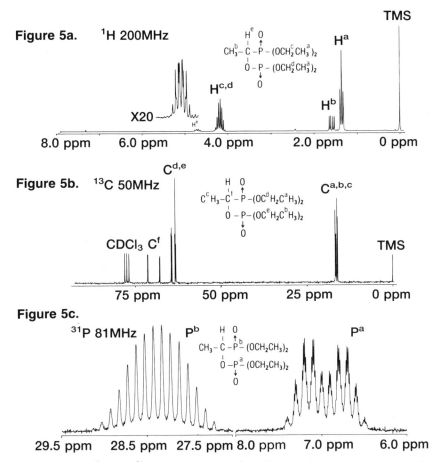

Figure 5. ¹H, ¹³C and ³¹P NMR spectra of an Organophosphorus compound.

the application of multinuclear NMR in identification of another organo-phosphorus compound. The ³¹P NMR (Fig. 5c) quickly establishes the presence of two different phosphorus atoms present as phosphate and phosphonate esters. Integration of the ¹H spectrum (Fig. 5a) reveals four types of protons present in the ratio 12:3:8:1. The ¹³C NMR spectrum (Fig. 5b) confirms the proposed structure with chemical shift information and C-P coupling constants. The multiline pattern in Figure 5c can be completely understood by first order analysis and gives P-H and P-P coupling constants which are confirmed by the ¹H spectrum.

Figure 6 is a ¹³C FT NMR spectrum obtained at 50 MHz on a 20% solution of an ethylene-vinyl acetate copolymer in 1 ml of chloroform-d. The spectrum was acquired in four hours. From this spectrum one can determine the percent vinyl acetate in the copolymer, the distribution of the vinyl

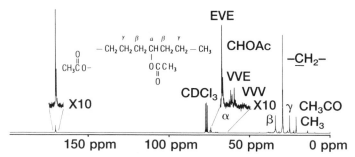

Figure 6. Ethylene-vinyl acetate copolymer, ^{13}C, 50 MHz.

acetate (random or block) and the presence and nature of branches in the polymer. The carbonyl carbon (172ppm) and the carbon α to the acetate (65–75ppm) are sensitive to triad structure. The α-carbon also shows pentad sensitivity and sensitivity to stereochemistry. (Triad and pentad units are three and five monomer units in sequence, respectively). The two peaks identified as due to a VVE triad are meso and dl 1,3-diacetates. The EVE:VVE:VVV ratio can be used to determine the polymer sequence distribution, which in this case is random.

Limitations

NMR Spectroscopy is a relatively insensitive technique when compared to other forms of spectroscopy. NMR Spectroscopy is also limited to those nuclei which possess a magnetic moment. Most elements have at least one stable isotope with a magnetic moment. However, in many cases the magnetically active isotopes are in low natural abundance, further limiting sensitivity. This problem can often be overcome by isotopic enrichment. Many materials are insoluble or poorly soluble in solvents suitable for high resolution NMR analysis further limiting sensitivity.

References

(1) Becker, E. D., High Resolution NMR; Academic Press Inc.: New York, 1980.

(2) Carrington, A.; McLachlan, A. D., Introduction to Magnetic Resonance; Harper & Row: New York, 1967.

(3) Fukushima, E.; Roeder, S. B. W., Experimental Pulse NMR; Addison-Wesley Publishing Company, Inc.: Reading, MA, 1981.

(4) Harris, R. K., Nuclear Magnetic Resonance Spectroscopy; Pitman Publishing Inc.: Marshfield, MA, 1983.

(5) Slichter, C. P., Principles of Magnetic Resonance; 2nd ed., Springer-Verlag: New York, 1978.

NUCLEAR MAGNETIC RESONANCE IN SOLIDS

Use

Solid state NMR is a field that encompasses many techniques which have proven useful to chemists for many years. Indeed, since the beginnings of NMR, solid state measurements, on well chosen systems (e.g. single crystals) have often provided structural information which could not be obtained with any other technique. Recently developed techniques, which utilize advances in probe design and electronics, have given chemists totally new and exciting techniques which provide composition and structural information on intractable materials. In particular, the development of high speed magic angle sample spinning (MASS) NMR, has opened a new methodology for studying polymers, organic compounds, and inorganic compounds in the solid state.

Sample

Current sample holders require $\sim\frac{1}{4}$ g of solid sample and at least this much material is preferred. Under some circumstances, e.g. studying relatively abundant nuclei, such as Na, P, or Al, smaller amounts of sample may lead to satisfactory spectra.

Principle

The basic principles of NMR in solution also apply to the NMR study of solids except that the molecules are not free to rotate and tumble. Molecular motion in the liquid state is responsible for the narrow NMR lines observed; when this motion is restricted as in the solid state, then the NMR lines become hopelessly broadened and relaxation times become very long. An important mechanism leading to line broadening is the spread of chemical shifts due to the wide variety of molecular orientations with respect to the magnetic field. Unlike liquids, the orientation of the molecules in the solid state are not averaged out by motion. This shift spread can be eliminated by studying single crystals, but this technique severely limits the applicability of the method. The spread of chemical shifts can, however, be removed by rotating the sample rapidly ($>$200,000 rpm) about an axis oriented at an angle of 54°, 44' (the magic angle) with respect to the magnetic field. This technique is known as Magic Angle Sample Spinning (MASS).

Another mechanism which leads to severe broadening is the interaction of the nuclei under study with their magnetic neighbors. In organic materials, with ^{13}C as the observed nucleus, neighboring protons are the principle source of this type of broadening. Strong proton decoupling is required to remove this interaction. In the case of liquids, RF energy fields of $\sim$4 KHz are normally used, whereas in solids ten times this amount of energy is required for decoupling. Strong proton decoupling works well for carbon-13 NMR since this nucleus is not abundant and no decoupling is required to remove the ^{13}C-^{13}C magnetic interactions (the occurrence of ^{13}C-^{13}C pairs is expected to be rare).

The strong proton decoupling field can also be used to increase nuclear relaxation by a technique called cross polarization (CP). By spin-locking the protons and carbons at a common frequency (Hartmann-Hahn condition) the carbon nuclei can be made to relax at the more rapid rate of the proton. This allows more rapid data acquisition and better sensitivity.

If CP, MASS and strong proton decoupling are used, then in favorable cases a ^{13}C spectrum approaching that obtained in the solution state can be obtained. As a general rule, linewidths (^{13}C) are on the order of one ppm (compared to liquids where the linewidths are a hundred times narrower), but narrower lines are sometimes obtained (0.1 ppm). With this degree of resolution, useful compositional and structural information can be obtained relatively quickly. This ability to extend the power of NMR to the solid state offers considerable promise for the study of insoluble materials, cross-linked polymers, and a variety of inorganic substances such as zeolites.

Applications

High resolution solution NMR has had a revolutionary impact on the study of large and small molecules over the past twenty-five years. With the development of techniques to handle materials in the solid state, NMR has recently been extended to the study of solid materials. A natural area of active study for solid NMR is the analysis of polymeric materials which are intractable in available solvents. The process of curing or cross-linking of polymers can now be studied as well as the final products of these processes. Recently, the ^{13}C NMR analysis of solid coal samples is yielding considerable insight into the aliphatic/aromatic ratios as well as the total carbon content. Solid state NMR also provides information on molecular motion occuring in the solid state and yields detailed information concerning, e.g., phenyl ring flips and other dynamic behavior. Amorphous and crystalline regions of a polymer can often be differentiated giving information concerning short-range ordering (X-ray diffraction depends on medium- to long-range ordering). In Figure 1 are shown the solution (a) and solid (b) state NMR spectra of a polymer precursor which clearly contrast the two methods. Peaks b and c are complex due to multiple conformations in the solid state. Peak g is broadened by the adjacent nitrogen quadrupole.

The use of MASS NMR for the study of a large variety of inorganic zeolites containing magnetically active ^{29}Si and ^{27}Al nuclei has received much

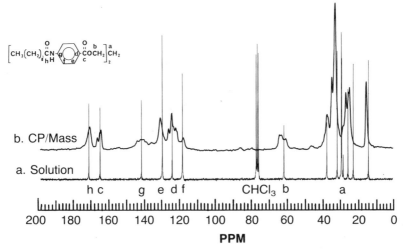

Figure 1. Solution (a) and solid (b) state NMR spectra of a polymer precursor.

attention recently. Solid state NMR gives a wealth of structural information concerning the framework of these aluminosilicates. Figure 2 is the ^{29}Si MASS spectrum of a zeolite, clearly showing the sensitivity of the method to the atoms attached to the SiO$_4$ tetrahedron.

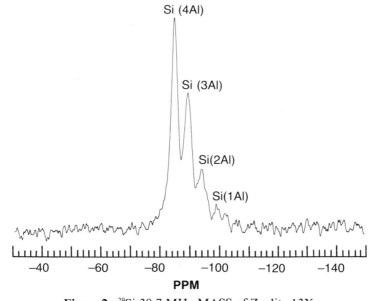

Figure 2. ^{29}Si 39.7 MHz MASS of Zeolite 13X.

Limitations

Solid-state NMR requires that the sample of interest contain a nucleus with a non-zero magnetic moment. In general, solid samples give much broader lines than are observed for similar samples in solution. This limits the resolution of the technique and the kind of structural information it can provide. Magic Angle Sample Spinning introduces spinning side bands into solid state NMR spectra which can be a nuisance in interpreting the spectra. In many cases, computer-aided lineshape analysis is needed for data interpretation.

References

(1) Fyfe, C. A., Solid State NMR for Chemists; CFE Press: Guelph, 1983.

(2) Gerstein, B. C.; Dybowski, C. R., Transient Techniques in NMR of Solids; Academic Press, Inc.: New York, 1985.

(3) Mehring, M., High Resolution NMR in Solids; 2nd ed., Springer-Verlag: New York, 1983.

2D-NUCLEAR MAGNETIC RESONANCE TECHNIQUES

Use

2D-NMR techniques provide great spectral simplification by spreading the conventional one-dimensional NMR spectrum in two independent frequency dimensions. This removes spectral overlap and facilitates spectral assignments. By choosing the proper pulse sequence, information on structure and conformation including through bond and through space interactions can be obtained. Correlation of ^{1}H and ^{13}C or ^{15}N spectra provides valuable connectivity information as well as improves sensitivity of ^{13}C and ^{15}N detection by up to two orders of magnitude. Quantitative information on atom interchange is also available. These techniques have been applied to both solution and solid samples and are in principle applicable to any NMR active nucleus.

Sample

2D techniques can be applied to the same solution NMR samples described in the previous sections. Because 2D experiments require more data acquisition than 1D experiments, larger samples are preferred.

Principle

In a 1D NMR experiment the FID can be acquired after a single pulse. In a 2D experiment a series of pulses are applied to the sample before acquiring the FID. In a typical 2D pulse sequence, preparation, evolution, mixing, and detection periods are separated by pulses of varying length as illustrated in Figure 1 for a 2D-NOE (Nuclear Overhauser Effect) experiment. Following the preparation period during which the nuclei are allowed to relax to their equilibrium magnetization, an initial pulse is followed by an evolution period (t_1), a second pulse, a mixing period, (Δ), a third pulse, and the detection period (t_2) during which the FID is recorded. By recording a series of FID's following varying evolution periods (t_1), a 2D time domain spectrum is obtained. Double Fourier transformation of this spectrum converts it to a 2D frequency domain spectrum which can be displayed as a stacked plot such as shown in Figure 2. While the stacked plot is impressive it is more convenient to present the data as a contour plot obtained by plotting a cross-section of the stacked plot as shown in Figure 3. The diagonal of the 2D plot represents the unperturbed resonances in the sample. During the evolution and mixing periods any interactions between non-equivalent nuclei generate off-diagonal components which are evident in the contour plot. Cross-peaks may result from exchange of nuclei, the Nuclear Overhauser Effect (NOE), or through scalar-coupling (J coupling). 2D experiments have been devised to exploit these and other interactions.

Applications

Two applications of 2D-NMR are illustrated in the accompanying figures. In Figure 2 is a stacked plot of ^{19}F homonuclear correlated experiment on a 40:60 ethylene:chlorotrifluoroethylene copolymer. In this experiment, the off diagonal peaks are produced by scalar coupling between fluorine atoms in close proximity to one another. From this experiment it was possible to work out the detailed microstructure of the polymer. 2D-NOE spectroscopy was used to assign the substitution pattern of the coumarin derivative illus-

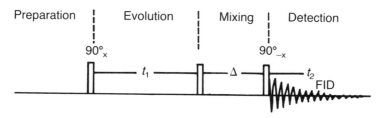

Figure 1. 2D NOE pulse sequence.

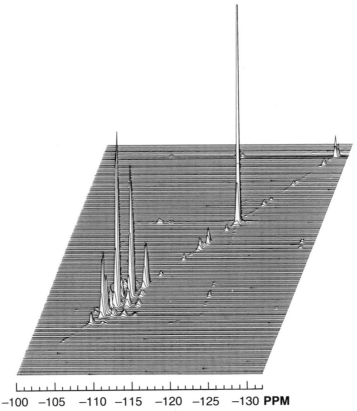

Figure 2. 2D-NMR spectrum of a 40:60 ethylene: chlorotrifluoroethylene copolymer.

trated in Figure 3. Cross-peaks are observed between the methyl resonance and protons a and b due to their close proximity to each other. Proton b also correlates with the benzylic methylene labeled γ. Proton c shows a cross-peak with the α- methylene group. The 2D-NOE experiment has allowed unambiguous assignment of all the resonances in this spectrum.

These two examples provide simple illustrations of only two 2D-NMR techniques which have found wide application in the structural characterization of small, medium and large molecules such as peptides, natural products, oligosaccharides, proteins, polynucleotides, and synthetic polymers. Other techniques include heteronuclear correlation experiments which have been used to identify directly coupled nuclei such as ^{1}H bonded to ^{13}C or ^{15}N. Using multiple quantum filters, it is possible to increase the sensitivity for low γ-nuclei such as ^{13}C or ^{15}N by observing directly attached nuclei such as ^{1}H which are more easily detected. Only a few of the 2D-NMR techniques which have been developed in the past several years have

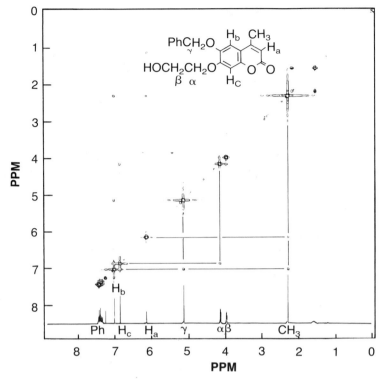

Figure 3. 2D NOE spectrum of a coumarin derivative.

been described. 2D-NMR has truly added a new dimension to NMR for the characterization of molecular structures.

Limitations

2D-NMR spectroscopy requires the acquisition of many 1D spectra before generating the 2D-spectrum. This requires a considerable increase in the time needed to generate a 2D spectrum. The multiple pulse sequences required to generate 2D spectra also place exacting demands on the NMR spectrometers. The large amounts of data and data processing required to generate a high resolution 2D spectrum place heavy demands on the computer capabilities of the spectrometer.

References

(1) Bax, A., 2D-Nuclear Magnetic Resonance in Liquids; Delft Univ. Press: Boston, 1982.

(2) Bax, A.; Lerner, L., Science, *232*, 1986, 960–967.
(3) Morris, G. A., Magnetic Resonance in Chemistry, *24*, 1987, 371–403.

ELECTRON PARAMAGNETIC RESONANCE SPECTROSCOPY

Use

Electron paramagnetic resonance spectroscopy (EPR), also known as electron spin resonance (ESR), gives information about materials with unpaired electrons, such as paramagnetic metal ion compounds, organic and inorganic radicals, semi-conductors and metals, and excited triplet state molecules. Useful structure and bonding information can often be obtained by this technique. With care, quantitative information on the concentration of paramagnetic molecules in a sample can be obtained.

Sample

Solids, liquids, or gases can be analyzed with EPR for paramagnetic species. Up to one gram of solid or liquid material can be studied, but normally much smaller amounts of material are needed due to the sensitivity of the technique. Approximately 1×10^{13} unpaired spins can be detected and the sample size is usually adjusted accordingly.

Principle

In most cases of chemical bonding the valence electrons form closed shells and the material is EPR inactive. However, when a normal bond is broken to leave an unpaired electron, or the molecules contain an odd number of electrons in their normal oxidation state, e.g., nitrous oxide (NO), then the material will be EPR active. An unpaired electron, either free or in the environment of a molecule or atom, possesses spin angular momentum (spin quantum numbers $+\frac{1}{2}$ and $-\frac{1}{2}$). The presence of an electric field and angular momentum results in each unpaired electron having a magnetic moment which can interact with an external magnetic field (H). The unpaired electrons will align themselves in either of two energy states. The energy difference between these two states is given by:

$$\Delta E = gBH$$

where g is the spectroscopic splitting factor and B is the Bohr magneton. The measurement of the energy difference (ΔE) is the basis of the electron paramagnetic resonance experiments.

The magnetic interaction between the electron and the nuclear spins in the molecule gives an EPR spectrum which will consist of more than one line. This interaction is called "hyperfine coupling" and is a measure of the electron spin density at the nucleus. Analysis of the intensities and frequency separations of the transitions make it possible to determine the nature of the paramagnetic site.

A schematic of an EPR spectrometer is given in Figure 1.

Applications

EPR spectroscopy is an important research tool in both chemistry and physics. The technique is amazingly sensitive, being able to detect molar concentrations of paramagnetic species in the range of 10^{-1} to as low as 10^{-12}M. It is extremely valuable for the examination of paramagnetic impurities or centers in solids or liquids. Some of the areas where EPR has been utilized for industrial applications follow.

I. Polymer and Organic Chemistry

EPR has been used extensively to study polymerization processes which occur by free radical mechanisms. Radical initiators, as well as the growing polymer radicals have been studied and information concerning mechanisms and rates of reaction have been obtained. Degradation processes in polymers often proceed by free radical initiators and many of these centers have been characterized by EPR. Typical studies might involve irradiation of the polymer with UV light followed by identification of the radiation damage center. Various operations such as machining, grinding, and

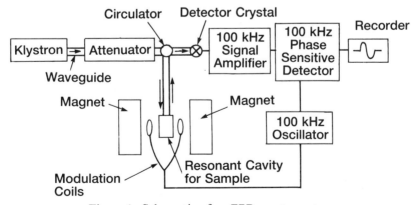

Figure 1. Schematic of an EPR spectrometer.

mechanical deformation lead to rupture in the polymer chain generating free radical ends which can be detected and quantified by EPR. Many antioxidants function by transforming highly reactive polymer radicals, formed e.g. by air oxidation, into more stable species, and EPR studies have been used to evaluate antioxidants.

A few organic compounds can exist as stable free radicals and often the EPR spectra are quite complex. Figure 2 shows the EPR solution spectrum of a derivative of N,N′ diphenyl-p-phenylenediamine. The complex splitting pattern observed is due to hyperfine interactions with the protons and nitrogens in the molecule.

II. Inorganic Chemistry

Many inorganic metals have oxidation states which are EPR active. The analysis of the EPR spectrum can often determine the oxidation state and the concentration of the metal present. Figure 3 shows the EPR spectra of a single crystal of alexandrite (Al_2BeO_4) containing chromium (III) ions oriented along the three principle axes (a,b,c) of the crystal. Both the concentration of Cr^{3+} and the crystallographic sites where the Cr ions substitute into the host lattice can be determined.

Figure 4 shows the first-derivative EPR spectrum of a nylon sample containing trace amounts (50μg/g) of Cu (II). The line shape and positions are completely characteristic for the cupric oxidation state.

Many reactions are catalyzed by transition metals and often free radical mechanisms are involved. For such reactions, EPR is an invaluable tool

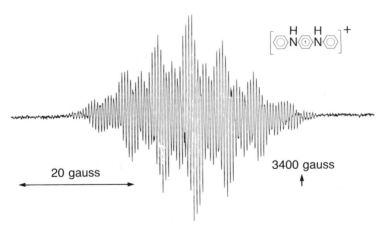

Figure 2. EPR solution spectrum of a derivative of N,N'-diphenyl-p-phenylenediamine.

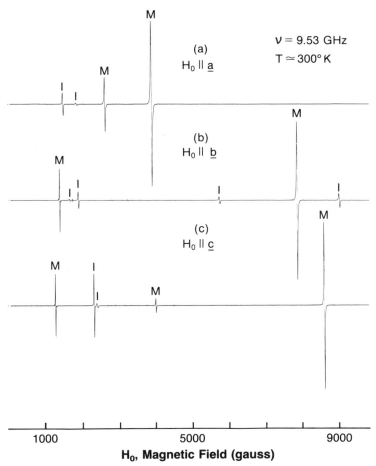

Figure 3. EPR spectra of a single crystal of alexandrite (Al_2BeO_4) containing Cr^{3+}.

for determining the mechanism and identifying the active site of the reaction.

Limitations

EPR spectroscopy is limited to those systems which possess unpaired electrons such as paramagnetic ions and free radicals, which are generally unstable. The amount of information obtained from EPR spectra while unique, is generally limited and requires computer simulations for proper interpretation. While EPR spectroscopy is very sensitive it is also difficult to quantify.

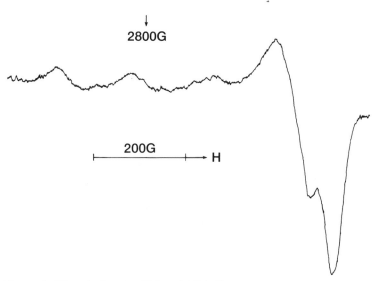

Figure 4. First-derivative X-band EPR Spectrum of Cu (II) in nylon.

References

(1) Bersohn, M.; Baird, J. C., An Introduction to Electron Paramagnetic Resonance; W. A. Benjamin Inc.: New York, 1966.

(2) Poole, C. P., Jr., Electron Spin Resonance, 2nd ed., A Comprehensive Treatise on Experimental Techniques; John Wiley & Sons: New York, 1983.

(3) Ranby, B.; Rabek, J. F., ESR Spectroscopy in Polymer Research; Springer-Verlag: New York, 1977.

Acknowledgment

Figure 1 is reprinted from: Kirk-Othmer, Encyclopedia of Chemical Technology; 3rd ed., John Wiley and Sons Inc.: New York, copyright© 1978; 669; with permission of John Wiley and Sons Inc.

Figure 3 is reprinted from: Forbes, C. E., J. Chem. Phys., 79, 1983, p. 2593; with permission of American Institute of Physics.

CHAPTER 3

Mass Spectrometry

R. Donald Sedgwick, David M. Hindenlang

LOW AND HIGH RESOLUTION MASS SPECTROMETRY

Use

Mass spectrometry can be applied to the quantitative analysis of organic, organometallic, inorganic and ionic compounds and materials including metals, and alloys. It is used both to confirm the presence of known compounds and to identify compounds of unknown structure.

Volatile or gaseous samples, usually organic compounds can be examined using electron impact (EI) or chemical ionization (CI) to give positive or negative ion mass spectra. Nonvolatile samples can be examined by fast atom bombardment (FAB) or thermospray ionization for inorganic and organic salts while plasma discharge ionization is used for metals and refractory materials.

A variety of variable temperature sample inlet systems are available for sample introduction. Complex mixtures can be separated on-line by either gas chromatography or liquid chromatography as appropriate.

Most instrumentation is computer controlled and incorporates extensive data acquisition, manipulation and presentation facilities to add both speed and reliability to the analyses.

Samples

Compounds in the molecular weight range of 2 to 3000 can be analyzed routinely with the possibility of obtaining some information up to 10000 amu.

While it is necessary to volatilize the sample for EI and CI modes of operation this is not a severe limitation for most non-ionic compounds since samples with vapor pressures of only 10^{-8} torr can be analyzed. Samples may be introduced at room temperature or temperatures up to 1000°C using heated probes which may be used to study thermal decomposition products of samples such as high molecular weight polymers. Volatile mixtures may be separated on-line by temperature programmable gas chromatography using a variety of packed or capillary columns.

Nonvolatile ionic compounds are admitted to the instrument in solution in an ionizing solvent system. For FAB ionization the solvent system must have low volatility, e.g. glycerol, while for thermospray ionization volatile solvents are used, e.g., water/methanol.

Metals and refractory materials can be admitted as solids for plasma discharge ionization.

In general one microgram of sample is sufficient for most purposes and it is not uncommon to detect subnanogram amounts of material ($<10^{-9}$g).

Principle

Mass spectrometry involves the production of gas-phase ions from a sample followed by their separation according to their mass-to-charge ratio (m/z) by a combination of electric and magnetic fields. The flux of ions at a particular mass is detected as an ion current. A mass spectrum is produced by measuring the ion currents corresponding to each value of m/z over a particular mass range. Depending on the ionization method selected mass spectra may be recorded as either positive or negative ion spectra.

The mass spectrum most usually includes a molecular ion whose m/z equals the monoisotopic molecular weight of the compound. Additionally the ionization method may induce fragmentation of the molecular ion to give ions of lower mass. These fragment ions are a source of chemical structural information for the molecular ion and its corresponding neutral structure. The partition of ions between molecular and fragmented species is determined by both the molecular structure and the ionization mode. EI spectra tend to have abundant fragment ions with attendant structural information. CI and FAB spectra emphasize the molecular weight information and commonly give stabilized molecular species formed by proton transfer ion-molecule reactions, $(M + H)^+$ and $(M - H)^-$. In these and other cases where fragmentation is not abundant, it can be induced by gas-phase collisional activation of a mass selected ion. The products of this collisionally induced dissociation can be analyzed in a second mass analyzer. Thus tandem mass spectrometry (MS-MS) may be used to define ionic decomposition pathways. Additionally unimolecular decompositions of ions can be defined and used as a valuable source of structural information by various modes of simultaneously scanning a tandem instrument. Details of this technique are given in the section on Tandem Mass Spectrometry.

A schematic of a double focusing magnetic mass spectrometer is given in Figure 1.

Applications

1. Nominal molecular weights can be obtained from low resolution mass spectra. High resolution can provide exact mass measurement accurate to 0.1 millimass unit and therefore yielding the corresponding unique atomic composition.

   For example, Figure 2 gives the computer printout from a high resolution mass spectrometry application. One peak (the parent ion) had a measured mass of 251.1543. Listed to its right are the calculated masses for the empirical formulae which are in a range of 30 millimass units of the measured peak; the empirical formula $C_{14}H_{21}NO_3$ is the best match that was consistent with other data. Similarly, a fragment peak (resulting from loss of a methyl group) had a measured mass of 236.1283, the best match being $C_{13}H_{18}NO_3$.

2. The molecular structure of a known compound can be confirmed and that of unknowns can be determined, especially when combined with other spectroscopic data.

   Known compounds may be identified automatically by computer matching of mass spectra with those in a data base. Structural features of unknown compounds can be deduced from an interpretation of the ionic fragmentation mechanisms seen in the mass spectrum. These deductions would normally be combined with IR, NMR and other spectroscopic data before a full structural assignment of an unknown would be made.

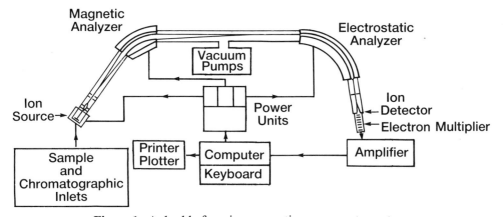

Figure 1. A double focusing magnetic mass spectrometer.

$$(H_3C)_3C \overset{\displaystyle OH}{\underset{\displaystyle NO_2}{\bigcirc}} C(CH_3)_3$$

ELEMENTAL COMPOSITION   DATA: 00515F # 25   BASE M/E: 55
05/19/80 13:52:00 + 6:40   CALI: 00515F # 25   RIC: 36800.
SAMPLE: CW YELLOW 00515
- (/S)
MIN INTEN: 72.   DEF: 30 MMU/100 AMU

MIN   53   0.00   0.   2.0
MAX   393   # 0   30.0 100 20.0 C200. H400. N4. O5

MASS	% RA	INTEN.	MMU	PPM	BONDS	FORMULA
208.1055	6.53	486.	-19.6	95	9.0	C16.H16
			-7.0	34	9.5	C15.H14.N
			5.4	26	10.0	C14.H12.N2
			18.0	87	10.5	C13.H10.N3
			16.7	80	10.0	C15.H12.O
			29.3	141	10.5	C14.H10.N.O
			-26.8	129	5.0	C10.H16.N4.O
			-28.1	136	4.5	C12.H18.N.O2
			-15.6	75	5.0	C11.H16.N2.O2
			-3.0	15	5.5	C10.H14.N3.O2
			9.5	46	6.0	C9.H12.N4.O2
			-4.3	21	5.0	C12.H16.O3
236.1283	19.66	1464.	-28.1	119	9.0	C18.H20
			-15.6	66	9.5	C17.H18.N
			-3.0	13	10.0	C16.H16.N2
			9.5	40	10.5	C15.H14.N3
			22.1	94	11.0	C14.H12.N4
			8.1	35	10.0	C17.H16.O
			20.7	88	10.5	C16.H14.N.O
			-24.1	102	5.0	C13.H20.N2.O2
			-11.5	49	5.5	C12.H18.N3.O2
			0.9	4	6.0	C11.H16.N4.O2
			-12.9	55	5.0	C14.H20.O3
			-0.3	1	5.5	C13.H18.N.O3
			12.2	52	6.0	C12.H16.N2.O3
			24.8	105	6.5	C11.H14.N3.O3
			23.4	99	6.0	C13.H16.O4
			16.2	69	2.0	C7.H16.N4.O5
251.1543	1.05	78.	-25.6	102	8.5	C19.H23
			-13.0	52	9.0	C18.H21.N
			-0.5	2	9.5	C17.H19.N2
			12.0	48	10.0	C16.H17.N3
			24.6	98	10.5	C15.H15.N4
			10.7	43	9.5	C18.H19.O
			23.3	93	10.0	C17.H17.N.O
			-21.6	86	4.5	C14.H23.N2.O2
			-9.0	36	5.0	C13.H21.N3.O2
			3.5	14	5.5	C12.H19.N4.O2
			-10.3	41	4.5	C15.H23.O3
			2.1	9	5.0	C14.H21.N.O3
			14.7	59	5.5	C13.H19.N2.O3
			27.3	109	6.0	C12.H17.N3.O3
			26.0	103	5.5	C14.H19.O4

Figure 2. High resolution mass spectrum computer printout.

3. Isotopic abundance can be measured and used to detect the presence of elements such as chlorine, bromine, sulfur, silicon, boron and other metallic elements with abundant and distinctive isotopic patterns.

The specific labeling of compounds with stable isotopes such as ^2H, ^{13}C, ^{15}N, and ^{18}O can be used to follow reaction mechanisms etc. and can be monitored by mass spectrometry.

4. Pyrolyses can be carried out prior to ionization inside the mass spectrometer or externally as in pyrolysis GC.

This is useful for characterizing materials which are nonvolatile or difficult to ionize directly. Thus, both linear and cross-linked polymers can be classified using pyrolysis-MS "fingerprints" as can many ionic compounds.

5. The sensitivity and specificity of mass spectrometry make it suitable for the analysis of trace-level organic compounds such as impurities in drinking water, wastewater, air and other gas samples.

Thus MS is used extensively in many EPA protocols for pollutant analyses for halocarbons, pesticides and general organics in environmental samples. Equally, trace level analyses of impurities in commercial samples of a very wide variety are carried out routinely.

6. Quantitative results can be obtained through the use of internal and external standards.

It is not uncommon to obtain linear concentration/response calibration curves over four orders of magnitude when EI and CI are used. Since ionization efficiency is not strongly dependent on chemical structure, MS is a useful "universal" detector system which can be relied on to give good semi-quantitative analytical data even in the total absence of calibration data.

7. The range of applicability of the gaseous ionization techniques (EI and CI) can be extended by forming volatile chemical derivatives from nonvolatile samples.

Limitations

Gaseous ionization techniques are not widely used for compounds with molecular weights over 1000. Volatility requirements usually rule out highly polar or ionic compounds of any molecular weight and all high polymers for examination by gaseous ionization methods.

The accuracy of high resolution mass measurement ($\sim$10 ppm) diminishes its usefulness as the mass increases. Since this represents an uncertainty of 5mmu at $m/z = 500$ the technique to determine atomic composition is only rarely useful above this mass.

References

(1) Burlingame, A. L.; Baillie, T. A.; Derrick, P. J., Anal. Chem. 58(5), 1986, 165R.

(2) Burlingame, A. L.; Whitney, J. O.; Russell, D. H., Anal. Chem., 56(5), 1984, 417R.

(3) Cooks, R. G.; Busch, K. L., Science, 222, 1983, 273.

(4) Watson, J. T., Introduction to Mass Spectrometry (2nd Ed), Raven Press: New York, 1985; pp 1–351.

CHROMATOGRAPHY—MASS SPECTROMETRY

Use

Complex volatile mixtures can be separated by GC and analyzed by on-line EI or CI mass spectrometry.

Applications

Both packed and capillary column GC may be used with only one restriction viz. the carrier gas used must be helium to facilitate its selective removal in the GC-MS interface.

Figure 1 shows the gas chromatogram obtained by plotting the variation of EI total ion current produced by a mass spectrometer accepting the eluant from a gas chromatograph.

Fast repetitive scanning and computerized data acquisition allow collection of more than sixteen hundred MS scans in a half-hour run of the GC.

Each peak represents a separated component in the mixture. The peak labelled 927 maximized at scan #927 of the mass spectrometer and its full mass spectrum can be retrieved from the raw data as shown in Figure 2.

This low resolution spectrum shows the correct fragmentation pattern, molecular mass and chlorine isotopic abundance patterns consistent with it being identified as tetrachloroethylene.

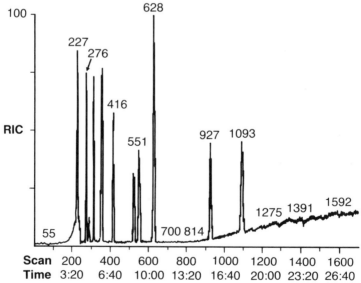

Figure 1. Reconstructed total ion chromatogram from a volatile mixture by GC-MS.

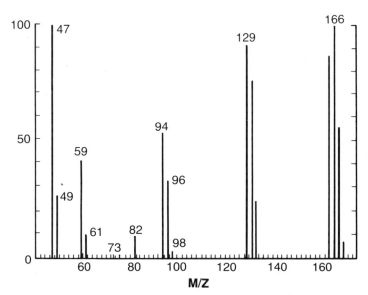

Figure 2. Electron impact mass spectrum of scan #927 in the chromatogram in Fig. 1.

The peak at 164 in the low resolution mass spectrum (Figure 2) can be mass measured at high resolution as 163.875$^+$. This mass is the best match for the atomic composition $^{12}C_2{}^{35}Cl_4{}^+$ confirming the assignment.

Quantitation can be obtained by reconstructing the ion current profile for a selected ion characteristic of the identified component. This is shown in Figure 3. The characteristic ion for tetrachloroethylene is m/z 164. The information above the peak indicates that m/z 164 maximized in scan #927 with an absolute intensity of 1652 and an absolute area of 15374. This area can be converted to a concentration in the original sample using internal and external standards. This type of data acquisition and reduction is typically used for analysis of organic Priority Pollutants in industrial wastewater.

Similarly complex nonvolatile mixtures can be separated by HPLC and analyzed by on-line thermospray mass spectrometry. This technique is complementary to GC-MS allowing mass spectra of ionic and highly polar compounds to be obtained when they are encountered in complex mixtures.

As with GC-MS the interface between the chromatograph and the MS ion source is a critical element needed to remove most of the solvent from the HPLC effluent.

In Thermospray ionization the HPLC effluent is sprayed through a heated nozzle into the ion source which is provided with a very high capacity vacuum pumping system. A volatile ionic buffer such as ammonium acetate is added to the mobile phase and this material serves to ionize the sample.

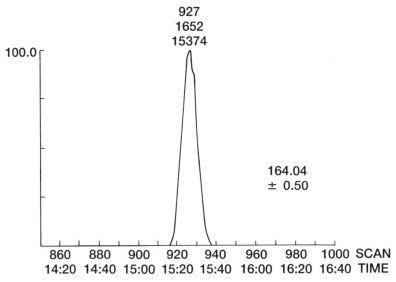

Figure 3. Selected single ion GC-MS trace for m/z 164.

The ionized sample molecules are desolvated in the heated spray and pass into the MS for analysis. In some systems ionization is achieved in the absence of an ionic buffer using an electrical discharge.

An alternative HPLC-MS system sprays the HPLC effluent onto a moving belt which passes through a series of heated vacuum locks on its path through a FAB ion source. The desolvated sample on the moving belt is ionized by xenon atom bombardment and the belt is then washed clean of sample before the HPLC effluent spray.

Both systems are compatible with normal and reverse phase HPLC with the maximum flow rates limited to <2ml per minute by the capacity of the desolvating pumping system.

Limitations

GC-MS is generally limited to volatile, thermally stable organic compounds with molecular weights below ~1000. This excludes most ionic compounds and highly polar compounds particularly molecules of biomedical significance. This group is usually amenable to HPLC-MS which is generally limited to non-volatile, polar compounds with molecular weights up to 5000.

GC conditions require the use of a helium carrier gas. HPLC conditions are limited by flow rate (<2ml per minute) and the use of volatile buffers in Thermospray. Mobile phases from 100% water to 100% acetonitrile can be used with full solvent programming.

References

(1) Covey, T. R.; Lee, E. D.; Bruins, A. P.; Henion, J. D., Anal. Chem., 58(14), 1986, 1451A.
(2) Holland, J. F.; Enke, C. G.; Allison, A.; Stults, J. T.; Pinkston, J. D.; Newcome, B.; Watson, J. T., Anal. Chem., 55(9), 1983, 997A.
(3) Lattimer, R. P.; Harris, R. E., Mass Spectrom. Rev., 4(3), 1985, 369.
(4) Rose, M. E., Specialist Periodical Reports—Mass Spectrom., 8, 1985, 210.

FAST ATOM BOMBARDMENT

Use

Nonvolatile, thermally labile, high molecular weight compounds can be examined using FAB ionization.

Sample

The sample is dissolved in a low volatility solvent of high polarity. This allows dissolution of ionic and highly polar involatile organic and inorganic compounds. The low volatility solvent can be glycerol, thioglycerol, polyethers, etc.

Principle

The sample in solution is mounted as a single droplet of volume $\sim 1\mu L$ on a metal target which is inserted into the evacuated ion source through a vacuum lock. Inside the ion source the surface of the liquid droplet is bombarded by a beam of energetic neutral atoms. Typically xenon atoms with kinetic energies of 10 keV are used, being formed in a collimated beam by charge exchange within a high pressure xenon ion source.

Secondary ions typical of both the sample and its supporting solvent are sputtered from the liquid surface and are analyzed in the mass spectrometer. Ionic organic compounds are frequently good surfactants in polar solvents and give good surface coverage of solvent surfaces even at low concentrations. Consequently surfactants give high sensitivity in FAB ionization and nanogram amounts of sample can be detected routinely in suitable cases.

The presentations of the sample in a liquid matrix means that a constantly refreshed surface is exposed to the sputtering atomic beam and radiation damage is minimized giving stable mass spectra. This may be contrasted with secondary ion mass spectrometry of organic solids which are typically unsuccessful due to the effects of radiation damage modifying the sample surface.

The use of a neutral atomic beam for bombardment gives equal efficiency for sputtering of both anions and cations. It also reduces undesirable electrical charging of insulating samples which can occur with ion bombardment and lead to the suppression of ion emission from the surface.

Applications

Compounds up to molecular weights of ~5000 can be examined and this includes many oligomeric compounds. Small proteins such as insulin can be analyzed intact by this technique with precise molecular weight determination possible from either the positive or negative ion FAB-MS as shown in Figure 1. This technique gives $(M + H)^+$ and $(M - H)^-$ ions as shown and the molecular formula $C_{254}H_{377}N_{65}O_{75}S_6$ for this fifty unit polypeptide gives an isotopic pattern commencing at mass 5729 with higher mass peaks arising from the distribution of stable isotopes in this formula.

This technique can be used to analyze high molecular weight biopolymers such as oligopeptides and oligonucleotides, most classes of antibiotics and synthetic surfactants of all types including carboxylates, sulfonates, polyethers and amines. Industrial compounds incorporating these ionic functional groups as well as phenolics are used as polymer additives, polymer stabilizers, finishes and processing aids and can be readily characterized. Inorganic compounds can also be analyzed as shown in Figure 1 where clusters of CsI were used to calculate the mass scale.

In Figure 2 is shown the positive ion FAB-MS of a commercial polyether showing the exact molecular weight-distribution. The material is a polyethylene glycol type material with oligomers containing different numbers of ethoxylate units of mass 44. The most abundant oligomer occurs at m/z = 590^+ with four lower and seven higher oligomers clearly visible. The ion at 590^+ is $(M + H)^+$ for the odd molecular weight of 589, since the ethoxylate has a nitrogen containing endgroup. It is evident that above m/z = 634^+ the molecules have mass defects which exceed 0.5 amu so that the displayed masses are rounded up to the next highest integer. Using the oligomeric distribution shown it is possible to calculate both the number-average and the weight-average molecular weight of the sample.

Both the positive ion and negative ion FAB-MS of a sulphonated azo dye are shown in Figure 3. The material is a sodium salt with a molecular weight of 496 thus the positive ion spectrum shows $(M + Na)^+$ at 519^+ while the negative ions show $(M - H)^-$ at 495^-. At lower masses in each ionization mode are seen structurally significant fragment ions which can be used to confirm the suspected molecular structure of the dye.

FAB-MS can also be used to identify surface adsorbed species and con-

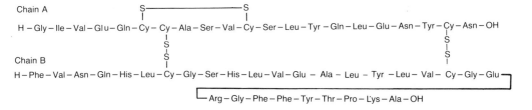

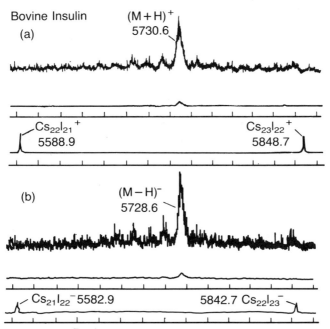

Figure 1. FAB-MS of the molecular ion region of bovine insulin using CsI to calibrate the mass scale.

taminants on metals and inorganic substrates with some limited depth profiling being achieved. Surface adsorbed layers can be identified and sputtered sequentially until the underlying substrate is identified. This is useful in studying corrosion and other surface effects.

Limitations

FAB-MS is generally limited to highly polar and ionic compounds which are soluble in a suitable polar solvent of low volatility. It is the method of choice for examination of most biomolecules, particularly polypeptides, polysaccharides and oligonucleotides. Choice of the solvent is critical in determining performance.

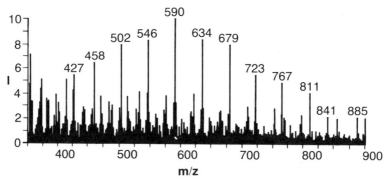

Figure 2. Molecular weight distribution of a commercial polyether sample by positive ion FAB-MS.

Sensitivity is usually lower than with other MS techniques but with fairly commonplace ability to detect nanogram quantities of material, FAB-MS nevertheless is one of the more sensitive methods of analysis.

Solution adsorption effects can cause selective sensitivity effects which make the application of FAB-MS for accurate quantitation of limited value.

High molecular weight compounds can be examined with the upper limit usually being set by the performance of the mass spectrometer. Compounds with molecular weights in excess of 5000 are frequently studied.

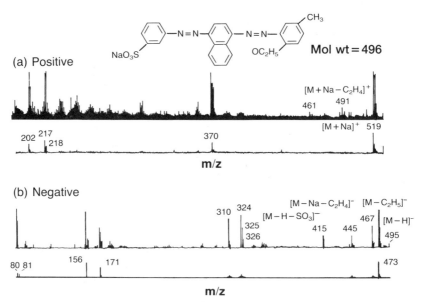

Figure 3. FAB-MS of a sulphonated azo dye.

References

(1) Barber, M.; Bardoli, R. S.; Elliott, G. J.; Sedgwick, R. D.; Tyler, A. N., Anal. Chem. 54, 1982, 645A.
(2) DePau, E., Mass Spectrom. Rev., 5(2), 1986, 191.
(3) Rinehart, K. L., Anal. Chem. Symp. Ser. 24, 1985, 119.
(4) Williams, D. H., ACS Symposium Ser., 291, 1985, 217.

TANDEM MASS SPECTROMETRY

Use

The mass spectrum of a mixture of compounds is the sum of the mass spectra of the pure components in the mixture, weighted according to their individual relative concentrations. Consequently it is very difficult to assign individual ions, either parent ions or fragment ions, to a particular component. Tandem mass spectrometry allows the recording of the mass spectrum of a single mass, formed when it is dissociated by collision after separation from all other ions and so provides clear linkages between ions in a complex spectrum. This allows the structures of individual ions to be deduced and hence leads to the structural assignment of their molecular precursors.

Principle

The "soft" ionization techniques such as CI and FAB ionization lead to the production of even electron "stable" ions which may be reluctant to fragment spontaneously. This gives their mass spectra good molecular weight information content but at the expense of molecular structure information which relies on fragmentation of the ions being recognized. The coupling together of two mass spectrometers (MS-MS) overcomes this disadvantage. The first MS is used to ionize and separate the ion whose structure is of interest. This separated "pure" ion beam is then focused into a collision chamber which causes collision-induced dissociation. These lower mass collision fragment ions are then analyzed in a second mass spectrometer and yield a spectrum characteristic of the parent ion's structure.

The tandem analyzers can be two quadrupole mass filters, a sector magnet followed by an electric sector (or vice versa) or other combinations.

The use of two quadrupole mass spectrometers separated by a quadrupole collision cell is referred to as a triple quadrupole mass spectrometer. The quadrupole collision cell conducts a wide mass range of ions and can be operated containing a high pressure of a collision gas. Ions transmitted by the first quadrupole MS are dissociated into a variety of fragment ions (daughters) which are transported efficiently into the final quadrupole MS. By varying the ways in which the first and final MS are scanned, it is possible to record all daughter ions of a single parent ion, all parents of a single daughter and all pairs of parents and daughters linked by a common neutral mass loss. Similar scans can be achieved using double focusing mass spectrometers using independent scan functions for the magnetic and electric sectors.

A drawing of a triple quadrupole mass spectrometer is given in Figure 1.

MS-MS scans on two element instruments give lower mass resolution than obtained for normal scans. Higher resolution can be achieved using more complex machines with three or four sectors which may also allow direct detection of parent, daughter, grand-daughter and even great-grand-daughter relationships.

Applications

The MS-MS technique is useful for the identification of components in complex mixtures where chromatographic separation is not possible. Figure 2 shows a negative ion FAB-MS of a sample of tetrabutyl ammonium phosphate which contains an "impurity" peak at m/z 157⁻. This relatively minor ion was separated from all other ions in a magnetic sector analyzer, dissociated by collision with helium gas, and the resulting fragment ions analyzed in an electric sector analyzer. This gave the spectrum in Figure 3 which identified the ion as the anion of benzene sulfonic acid. This identification was confirmed by accurate mass measurement showing mass 157⁻ to be exactly 156.992 corresponding with $C_6H_5SO_3^-$.

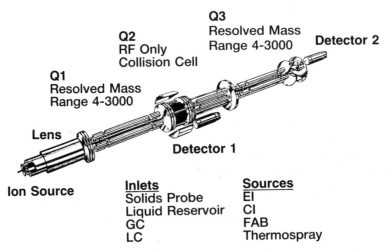

Figure 1. A triple quadrupole mass spectrometer for MS-MS applications.

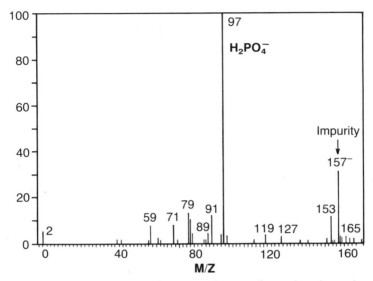

Figure 2. Negative ion FAB-MS of tetrabutylammonium phosphate showing an impurity peak at 157$^-$.

Limitations

MS-MS applications may be sensitivity limited due to inefficiencies in the collision processes and collection of the dissociation products. The effect of the relatively low response in the second MS on overall sensitivity is mitigated by excellent signal to noise characteristics. In some cases these competing effects may combine to produce an overall sensitivity gain.

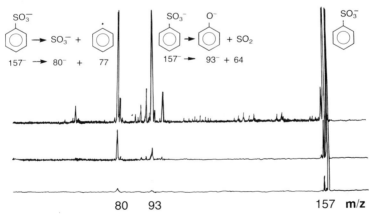

Figure 3. Collisionally induced dissociation mass spectrum of impurity peak at 157$^-$.

For standard instrumentation some results may be ambiguous due to attainment of insufficient mass resolution in the MS-MS mode.

References

(1) Chapman, J. R., Practical Organic Mass Spectrometry; John Wiley and Sons: New York, NY, 1985; pp 137–162.

(2) Crow, F. W.; Tomer, K. B.; Gross, M. L., Mass Spectrom. Rev., 2, 1983, 47.

(3) Johnson, J. V.; Yost, R. A., Anal. Chem., 57, 1985, 758A.

(4) McLafferty, F. W., ed.; Tandem Mass Spectrometry; John Wiley and Sons: New York, NY, 1983.

Acknowledgment

Figure 1 is reprinted with permission of VG Mass Lab., Manchester, England.

CHAPTER 4

Chromatography

James M. Hanrahan, Mina K. Gabriel, Richard J. Williams, Milton E. McDonnell

GAS CHROMATOGRAPHY

Use

Gas chromatography (GC) is used for qualitative and quantitative analysis of complex mixtures of gases, liquids and some solids. It is usually the first analytical choice for quantitation of mixtures of the more volatile organic compounds.

Sample

Gases, liquids and volatile solids can be analyzed. Actual samples are of the order of one to ten microliters, though larger samples can be accommodated. Sensitivities of many components are sufficient to allow detection at the ppm and even the ppb level.

Principle

A vaporized sample is transported by an inert carrier gas through a column of suitable length to provide the desired separation. The column is usually packed with a partitioning liquid coated on the surface of an inert solid support. Wall coated open tubular (capillary) fused silica columns have recently achieved wide acceptance. Active uncoated supports are also used. The various components of the vapor mixture are separated as a result of their relative vapor pressures and their relative affinities for the liquid partitioning phase. Normally, more volatile components emerge in shorter times than those with lower vapor pressures, though this order can be reversed by suitable choice of the liquid partitioning phase. The emergence of each component is signaled by a suitable detector and is displayed on a recording device. Thermal conductivity and flame ionization detectors are used in most cases. Electron capture, flame photometric, nitrogen/phosphorus, photoionization, and electrolytic conductivity (Hall) detectors are also available. In the more modern instrumentation, chromatographic operation can be handled by a micro-processor, thereby bringing considerable automation to analysis by gas chromatography.

Among the detector types mentioned above, thermal conductivity is applicable for quantitation of major components; flame ionization is used for general high sensitivity to trace amounts; electron capture provides for very high sensitivity to electronegative groups; and other specialized detectors (nitrogen/phosphorus, flame photometric, Hall, etc.) are particularly sensitive to elements such as sulfur, phosphorus, nitrogen and chlorine.

A schematic of a gas chromatograph is given in Figure 1.

Applications

Gas chromatography is the single most important tool for relatively rapid separation of volatile complex mixtures. This technique may be used in conjunction with such structure-determining instruments as infrared and mass spectrometry by direct interface, and with nuclear magnetic resonance through trapping of cuts for subsequent analysis. The result in many cases is complete characterization of a complex mixture in a relatively short time.

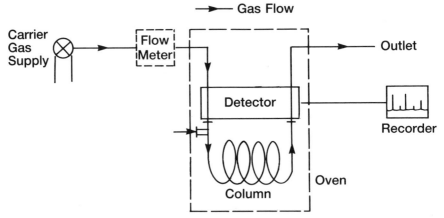

Figure 1. Schematic of a gas chromatograph.

Some specific applications are as follows:

1. Detailed characterization of complex mixtures such as alkylbenzenes and mixed aromatic-aliphatic plasticizer systems can be accomplished on capillary columns using flame ionization detection.
2. Pyrolysis studies on polymers in the 500–1200°C range can lead to information on branching, isomer content and other structural features.
3. Dual detection systems can be utilized with various combinations of flame, thermal conductivity and electron capture detectors for example, to provide more specific qualitative information.
4. Specific detection systems provide enhanced sensitivity to certain trace moieties while excluding or minimizing the response to extraneous components. Detection of 1) chlorinated hydrocarbons in drinking water, 2) extracted monomers in food-contacted plastic containers, and 3) catalyst-poisoning trace sulfur compounds in organic matrices are but a few applications of specific detectors.
5. Improved column technology such as development of liquid crystal phases for the separation of polynuclear aromatics and chiral center liquid phases for the resolution of d- and l-optical isomers have broadened the applications' scope.
6. Separation of complex "non-volatile" mixtures can be obtained by formation of volatile derivatives. For example, it is not possible to pass all types of polyamide hydrolysate products through gas chromatographic systems due to their low volatility. However, if the dried hydrolysates are treated with the dialkyl acetals of N,N-dimethylformamide the resulting products become volatile and stable enough for chromatographic purposes. An example of the separation of various polyamide hydrolysates is shown in Figure 2.

   Another example of gas chromatographic analysis of "non-volatile" compounds is given in Figure 3. The acid groups of "non-volatile" amino acids are first esterified. Trifluoracetic anhydride is then reacted with the amino groups to give products which are sufficiently volatile for gas chromatographic analysis. Figure 3 shows the separation achieved on fifteen amino acids utilizing this derivitization technique. Incorporation of a known amount of n-butyl stearate as an internal standard allows quantitative calculations of the individual amino acids.
7. Although capillary (Golay) columns appeared early in the history of gas-liquid chromatography, the recent development of the rugged and highly efficient fused silica capillaries has brought capillary chromatography to a position of dependable, everyday utility over a wide range of application. The analysis of highly complex cresylic and tar acid mixtures is an example of a sample type whose components can be successfully resolved only by capillary procedures. Lately the wide bore (e.g., 0.5mm) high

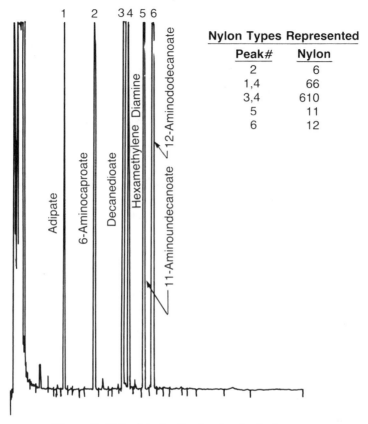

Figure 2. GC of derivatized nylon hydrolysates.

capacity, fused silica columns with bonded liquid phases are replacing conventional packed columns in everyday use.

Limitations

Non-volatile compounds cannot be analyzed unless pyrolysis or derivatization converts them to a condition amenable to chromatography. Supercritical fluid gas chromatography (SFC), however, allows the analysis of many compounds of limited or nil volatility. The analysis of thermally labile compounds is also precluded but again SFC might apply in these instances. Fixed gases do not respond to thermal ionization detectors but are detected by the less sensitive thermal conductivity detectors.

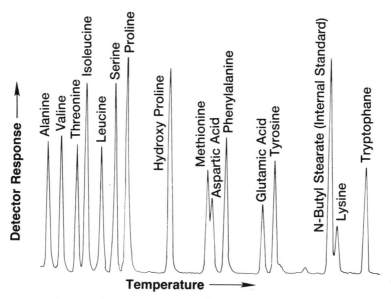

Figure 3. GC of *n*-Butyl esters of trifluoroacetylated amino acids.

References

(1) Clement, R. E.; Onuska, F. I.; Yang, F. J.; Eiceman, G. A.; Hill, H. H. Jr., Anal. Chem., 58, 1986, 321R–335R.

(2) Jennings, W. G.; Rapp, A., Sample Preparation for GC Analysis; Huethig Publishing: Mamaroneck, NY, 1983.

(3) Nikelly, J. G., Advances in Capillary Chromatography; Huethig Publishing: Mamoroneck, NY, 1986.

Acknowledgment

Figure 1 is reprinted from: Kirk-Othmer, Encyclopedia of Chemical Technology, 3rd ed., John Wiley and Sons Inc: New York, copyright© 1978; 600, with permission of John Wiley and Sons Inc.

LIQUID CHROMATOGRAPHY

Use

Liquid chromatography (LC) is used for qualitative and quantitative analysis of complex mixtures of liquids and soluble solids. Single components can be isolated from a mixture for structural analysis. Preparative liquid chromatography has advanced to the point where isolation of gram quantities of highly pure materials is straight-forward.

Sample

Solubility in a suitable solvent is the only criterion a sample must meet for liquid chromatographic analysis. Most analyses are done at ambient temperatures and neutral pH. Sensitivities in the ppm and even ppb range can be achieved depending on the type of detection employed and the nature of the sample being analyzed. Most analyses are done with a total sample in a microgram to milligram range. Many of these analyses can be modified to preparative scale so that highly pure gram quantitities of compounds can be obtained.

Principle

Separation of the components is achieved by the relative degree to which each component is associated with the solid (stationary) phase while being transported in the liquid (mobile) phase. In the earliest form of liquid chromatography, association to the solid phase is by adsorption, the common solid phase being silica or alumina. This type of chromatographic separation is still in common use. In current practice, the most widely used solid phase consists of a support medium to which is bonded, either physically or chemically, a material in which the solute species are soluble. The degree of partition between this immobile phase and the mobile solvent phase produces the separation.

LC (as compared to GC) has the advantage that the mobile phase can be chosen from a wide variety of solvents or solvent mixtures, thus adding considerably to flexibility. Also, the chromatographic process is conducted at ambient temperatures and (usually) at neutral pH; these mild conditions minimize any possible chemical reaction or degradation during analysis. On the other hand, LC, suffers by comparison to GC in that there is no universal LC detector.

In order to achieve separation and hence analysis in reasonable time, the sample is applied to a chromatographic column under operating conditions. A solution of the sample is injected onto the head of the column by means of a multi-port valve at the operating pressure of 7×10^6 to 30×10^6 Pa (1000–4000 PSI). The components are separated and eluted from the column where they are sensed by the detection system. Ultraviolet absorption using monochromatic radiation is a typical detection method. Photodiode-array technology coupled with powerful computers allows for rapid access to a three dimensional matrix of analytical information; absorption over the full wavelength range and time. For samples containing non-ultraviolet absorbing components, refractive index, or more recently, flame ionization detectors are used. The combination of a flowing system, injection of the sample under pressure, a specialized column in which the stationary phase consists of tailored molecules bonded to a substrate, and a flexible detection system—is referred to as High Performance Liquid Chromatography (HPLC). In Figure 1 is given a schematic of a high pressure liquid chromatograph.

A further refinement in HPLC is the use of solvent programming or gradient elution, in which the composition of the eluting solvent is changed in a stepwise or continuous manner. This has the advantage of yielding slower elution of the earlier peaks but faster elution of the tenaciously retained components. The process is analogous to temperature programming in gas chromatography.

Two of the earlier forms of LC are thin layer chromatography (TLC) and paper chromatography (PC). In TLC the chromatographic media is coated onto a glass plate. TLC can be used for both adsorption and partition chromatography, depending on the chromatographic media and solvent system employed. PC is used mainly for partition chromatography where the paper

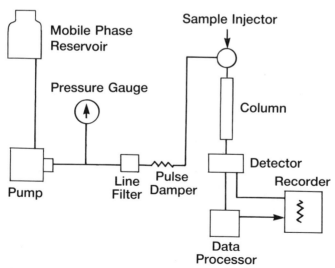

Figure 1. Schematic of a high pressure liquid chromatograph.

acts as a support for the stationary phase. In both techniques the sample is spotted as a solution at one end of the plate or paper. The solvent is evaporated and the plate or paper is placed in a pre-equilibrated chamber. The developing solvent is allowed to migrate up the plate by capillary action (TLC) or flow down the paper by gravity. Once the solvent has reached a suitable level the plate or paper is removed from the chamber and dried. The separated sample compounds can then be analyzed by a variety of methods. In many cases a specific spray reagent can be used. These methods of chromatography present a visual display of the separation and are often useful for diagnostic purposes in developing an HPLC analysis.

Applications

Liquid chromatography, because of its great flexibility, can be used for the analysis of over eighty percent of all known organic compounds. By modifying the separation method sufficient purified material can be isolated for subsequent study by infrared, mass spectrometry and NMR spectroscopy. Alternatively, direct interfacing of liquid chromatography (HPLC) to mass spectrometry and Fourier-transform infrared instrumentation is now possible. The results in many cases are complete characterizations of complex mixtures.

Some specific applications are:

1. Figure 2 shows the 15 minute analysis of a mixture of dinucleotides and nucleoside phosphates accomplished by ion-pair reverse-phase high performance liquid chromatography using an ultra-violet detector.

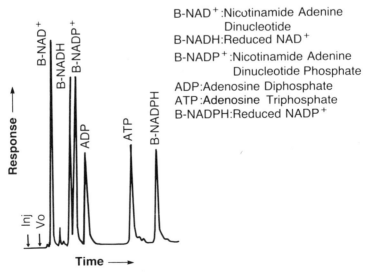

B-NAD$^+$:Nicotinamide Adenine
        Dinucleotide
B-NADH:Reduced NAD$^+$
B-NADP$^+$:Nicotinamide Adenine
        Dinucleotide Phosphate
ADP:Adenosine Diphosphate
ATP:Adenosine Triphosphate
B-NADPH:Reduced NADP$^+$

Figure 2. HPLC of nucleotide mixture.

2. Figure 3 shows the chromatogram of a complex mixture of polynuclear aromatic hydrocarbons obtained by partition chromatography. The detector in this case is a fluorescence detector, which is specific to compounds of this type and provides a high degree of sensitivity.
3. The analysis of two types of sulfonic acids in a solvent mixture was accomplished by partition chromatography, using an ultra-violet detector. The analysis can be done in 4 minutes; the chromatogram is shown in Figure 4.

Limitations

Gas chromatography generally provides fast and reliable separation of only 20% of known organic compounds without prior chemical modification of the sample. However, many samples simply cannot be handled by GC, either because they are insufficiently volatile and cannot pass through the column, or they are thermally unstable and decompose under conditions

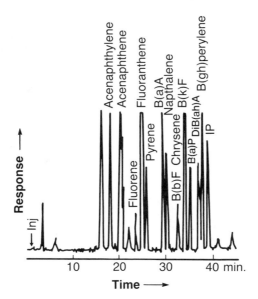

B(a)A: Benzo (a) Anthracene
B(b)F: Benzo (b) Fluoranthene
B(k)F: Benzo (k) Fluoranthene
B(a)P: Benzo (a) Pyrene
DiB(ah)A: Dibenzo (a,h)
            Anthracene
B(gh)Perylene: Benzo (gh)
            Perylene
IP: Indeno (1,2,3-cd) Pyrene

Figure 3. HPLC of PNA mixture.

Sulfonic Acids

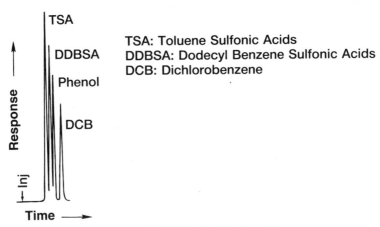

TSA: Toluene Sulfonic Acids
DDBSA: Dodecyl Benzene Sulfonic Acids
DCB: Dichlorobenzene

Figure 4. HPLC of sulfonic acids.

of separation. LC, on the other hand is not limited by sample volatility or thermal stability, and thus it is ideally suited for the separation of macromolecules and ionic species of biomedical interest and labile natural products. Samples, however, must be soluble in a suitable solvent to be amenable to analysis by LC. Although LC offers a number of unique detectors that so far found limited application in GC, it suffers from the lack of a universal detector.

References

(1) Krstulovic, A. M.; Brown, P. R., Reverse-Phase High-Performance Liquid Chromatography, Theory, Practice and Biomedical Applications; John Wiley and Sons, Inc.: New York, 1982.

(2) Pryde, A.; Gilbert, M. T., Applications of High-Performance Liquid Chromatography; Halsted Press, John Wiley and Sons, Inc.: New York, 1979.

(3) Snyder, L. R.; Kirland, J. J., Introduction to Modern Liquid Chromatography; 2nd edition, John Wiley and Sons, Inc.: New York, 1979.

Acknowledgment

Figure 1 is reprinted from reference 3, chapter 3 p. 86. copyright© 1979; with permission of John Wiley and Sons, Inc.

SUPERCRITICAL FLUID CHROMATOGRAPHY

Use

Supercritical fluid chromatography (SFC) is used to separate and quantitate mixtures of liquids and soluble solids as in liquid chromatography. It can be used to isolate individual components from mixtures in quantities sufficient for structural analysis by spectroscopic methods.

Sample

The liquid or solid mixture must be soluble in a solvent for dilution and injection purposes and must also be soluble in the mobile phase which is a supercritical fluid viz a "gas" above its critical pressure and above its critical temperature. Most analyses are carried out at room temperature or slightly higher but rarely above 100°C. Sensitivities in the ppm range or better are achievable depending on the method of detection used and the chemical nature of the sample. Total sample sizes are usually in the microgram to milligram range.

Principle

Components in a mixture are separated by their partition between the mobile phase and a stationary phase which is a component of a column. The mobile phase flows through the column and transports mixture components at variable rates so that they are eluted at different times. This is exactly the same as in liquid chromatography or gas chromatography except for the physical state of the mobile phase. Some of the supercritical fluid mobile phases are carbon dioxide, ammonia, monochlorodifluoromethane or dichlorodifluoromethane. The fluids may also be spiked with low levels of organic solvents such as alcohols and ethers in order to modify the elution characteristics of the mobile phases. The stationary phase may be coated on a solid support which is packed into a column or may be coated or bonded to the inner surface of a capillary column.

SFC clearly lies at the interface between GC and LC. The range of mobile phase characteristics is much more restricted than in LC but by using pumping technology similar to that of LC it is possible to modify the mobile phase characteristics by pressure programming. Pressure programming in the range 100–300 atm results in a fluid density change from 0.3 to 0.8 g cm^{-3} giving a mobile phase whose diffusivity and viscosity is intermediate between normal gases and liquids. Thus starting at relatively low pressures it is possible to achieve efficient and rapid separation of many compounds usually separated by GC. At higher pressures the system becomes very similar to LC and is efficient for compounds at or above the volatility boundary of GC and probably amenable to LC. The technique allows analysis in a reasonable time for those samples which are of too wide a volatility range, usually associated with a wide range of molecular weights, to be efficiently separated by either GC or LC.

Pressure programming using pure gases as mobile phases using capillary columns allows the use of the flame ionization detector (FID). This has the advantage of wide applicability to organic compounds and a sensitivity associated with conventional GC. The use of doped mobile phases has a dramatic effect on the efficiency of separation of polar materials but usually precludes use of the FID. These systems are commonly used with ultraviolet or refractive index detectors which are widely used in LC.

The fact that the mobile phase becomes gaseous when the pressure is reduced makes SFC a technique which can be interfaced to other spectroscopic techniques using well developed technology from GC.

A block diagram of a supercritical fluid chromatograph is given in Figure 1.

Applications

Supercritical fluid chromatography has advantages over both GC and HPLC in certain analytical applications. However, SFC will not supplant either of these classical forms of chromatography. Rather, it supplements the classical forms and provides another analytical tool to increase the analyst's ability to solve difficult analytical problems. SFC does not suffer from volatility limitations as GC does. For example, Figure 2 is a capillary SFC profile of a mixture of relatively polar polypropylene glycol oligomers hav-

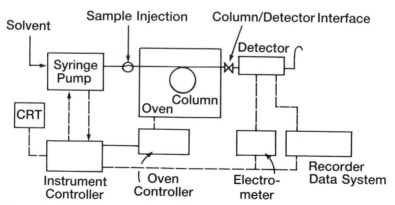

Figure 1. Block diagram of a capillary supercritical fluid chromatograph.

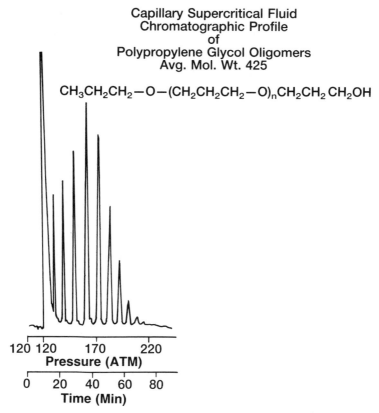

Figure 2. SFC separation of polypropylene glycol oligomers.

ing an average molecular weight of 425 amu. Analysis of these underivatized oligomers by GC is not possible due to the inability of GC to elute all of the oligomers at reasonable operating temperatures.

Capillary column SFC is compatible with many types of conventional GC and HPLC detectors, such as flame ionization detection (FID), nitrogen/ phosphorus detection (NPD), Fourier transform infrared spectroscopy (FTIR), mass spectrometry (MS), ultraviolet detection (UV), fluorescence detection, and others. As seen in Figure 3 SFC can be routinely used with a reliable universal detector such as FID, which HPLC cannot. The same figure also illustrates that SFC is compatible with other selective and element specific flame detectors such as the traditional NPD.

SFC methods for the separation of complex mixtures of polycyclic aromatic hydrocarbons from carbon black extracts by employing ultraviolet detection have been developed as illustrated in Figure 4. The combination of capillary SFC with on-the-fly scanning fluorescence detection created an

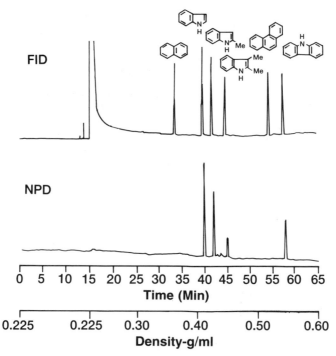

Figure 3. SFC separation of two polycyclic aromatic hydrocarbons and four nitrogen heterocycles. Two separate analyses were performed; one using an FID and a second using an NPD. Note the excellent selectivity of the NPD.

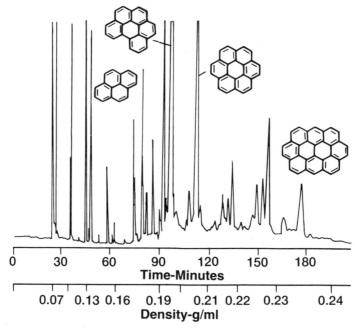

Figure 4. SFC separation of a carbon black extract.

extremely powerful analytical tool providing qualitative and quantitative information on the solutes and displaying sensitivity for some analytes in amounts as low as 50 pg as shown in Figure 5.

Limitations

SFC has several characteristics which make it useful for the study of wide molecular weight range mixtures and for situations which demand repeated high speed analysis of high molecular weight or thermally labile samples. The development of efficient, easy to use instrumentation will help propagate this analytical method. One aspect which needs to be further explored and demonstrated is the determination of the types of samples for which SFC is best suited and can handle. The narrower choices of mobile phases as compared to liquid chromatography put some restrictions on the sample types which can be conveniently handled. This also points out the need for more research on additional mobile phases, and the mixing or doping of mobile phases with small percentages of modifiers to achieve special effects. The use and further developments of insoluble stationary phases will be

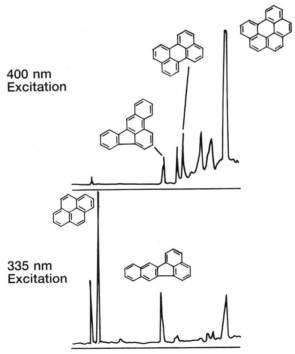

Figure 5. Selective fluorescence wavelength reconstructions of a SFC separation of a carbon black extract.

beneficial for providing better selectivity and allowing mobile phase conditions which normally wash other stationary phases out of the column.

References

(1) Peaden, P. A.; Lee, M. L., J. Liquid Chromatography, 5, 1982, pp 179–221.
(2) White, C. M.; Houck, R. K., J. High Resolution Chromatography and Chromatography Communications, 9, 1986, pp 4–17.

Acknowledgments

Figure 1 and 2 are reprinted from: White, C. M.; Houck, R. K., Journal of High Resolution Chromatography and Chromatography Communications, 9, 1986, 4; with permission of Suprex Corp. and ©Dr. Alfred Huethig Publishers, Heidelberg.

Figure 3 is reprinted from: Fjeldsted, R. C.; Kong, R. C.; Lee, M. L., J. Chromatogr., 279, 1983, 449; with permission of Elsevier Science Publishers B.V., Amsterdam.

Figure 4 is reprinted from: Jackson, W. P.; Richter, B. E.; Fjeldsted, J. C.; Kong, R. C.; Lee, M. L., High Resolution Supercritical Fluid Chromatography, Chapter in Ultra High Resolution Chromatography, ACS Symposium Series, Washington, D.C., 1984; with permission of the American Chemical Society.

Figure 5 is reprinted from: Fjeldsted, J. C.; Richter, B. E.; Jackson, W. P.; Lee, M. L., J. Chromatogr., 279, 1983, 423; with permission of Elsevier Science Publishers B.V., Amsterdam.

ION CHROMATOGRAPHY

Use

Ion chromatography (IC) is used for qualitative and quantitative analysis of ionic species in complex matrices.

Sample

Liquids and solids are analyzed after dissolution in deionized water. Gases which produce ionic species in water can be analyzed after absorption in an appropriate medium. Detection limits, under normal operating conditions, are in the sub-ppm to ppm range. The use of a concentrator column can lower the detection limits to the 1–10 ppb range.

Principle

Ionic species are separated on low capacity pellicular ion exchange columns using either sodium bicarbonate/sodium carbonate (anion separation) or dilute nitric acid (cation separation) eluents. Either a suppression column or fiber suppression system is used to decrease the high conductance background of the eluent. The suppression column consists of a high capacity ion exchange resin opposite in type to that of the separator column. For example, in anion analysis, a cation exchange $(Resin-SO_3^-H^+)$ column is used to suppress the eluent:

$$Resin -SO_3^-H^+ + NaHCO_3/Na_2CO_3 \rightarrow Resin-SO_3^- Na^+ + H_2CO_3$$

while the separated anions are converted to their corresponding acids:

$$Resin -SO_3^-H^+ + NaCl \rightarrow Resin-SO_3^- Na^+ + HCl$$
$$Resin -SO_3^-H^+ + NaNO_3 \rightarrow Resin-SO_3^- Na^+ + HNO_3$$

The main disadvantage of this approach is that the suppressor column must be regenerated with acid ($1N\ H_2SO_4$) after about eight hours of use. With a fiber suppression system, the sodium bicarbonate/sodium carbonate eluent is passed down the core of a hollow fiber cation exchange membrane while a counterflowing stream of acid on the outside continuously supplies hydronium ion (H_3O^+) to exchange for sodium ions across the membranes. Anions are prevented from passing through the hollow fiber by ion exclusion. From the suppressor column/fiber suppression system the separated anions flow into an electrical conductivity detector. UV/Vis or amperometric detectors are commonly used as auxiliary detection methods. Analysis of cations is exactly analogous but uses dilute nitric acid as the eluent, a low capacity cation exchange column for separation, and a high capacity anion exchange column or fiber suppression system for suppression of the eluent ions.

Transition metals can be separated on a low capacity pellicular ion exchange resin (either cationic or anionic) using complexing agents, such as oxalic, citric, tartaric, and isobutyric acids as eluents. Selectivity is dependent upon the metal, oxidation state, complexing agent and pH. After separation, the metals enter a postcolumn reactor which adds a reagent that selectively complexes with the metal ions to yield complexes that commonly absorb in the visible region. The postcolumn reactor effluent is monitored using a UV/Vis detector. This method has sensitivity comparable to and in some cases superior to graphite furnace atomic absorption spectroscopy.

Organic acids can be separated by ion-exclusion chomatography (ICE-ion chromatography exclusion). In ICE, the separator is a large high capacity cation exchange resin in the hydrogen form. Retention of the acids is controlled by the extent to which the acid is ionized in solution:

$$HA \rightleftharpoons H^+ + A^-$$

The neutral or unionized form of the acids undergoes partition or reverse phase chromatography in the cation exchange column. Selectivity depends mainly on the pk_a value (negative logarithm of the acidic dissociation constant) of the organic acids and the pH of the eluent. Dilute acids (usually HCl) are used as eluents. The eluent is suppressed using a disposable cation exchange column in the silver form:

$$Resin^-Ag^+ + HCl \rightarrow Resin^- H^+ + AgCl(s)$$

A fiber suppressor compatible with ICE analysis has also been developed. The schematic of an ion chromatograph is given in Figure 1.

Applications

IC allows the separation and detection of trace ionic species such as F^-, Br^-, NO_3^-, SO_3^{-2}, SO_4^{-2}, PO_4^{-3}, AsO_4^{-3}, SeO_4^{-2}, the alkali and alkaline earth metals, transition and lanthanide metals, ammonium and low molecular weight amines, quaternary ammonium compounds, weak organic acids (formic, acetic, etc.), aliphatic and aromatic sulfonic acids, carbohydrates, and amino acids. Some applications are as follows:

1. Determination of acid concentration in pickling baths: HF, HCl, HNO_3 and H_2SO_4—See Figure 2.
2. Determination of ionic species (both major and minor) in plating baths.

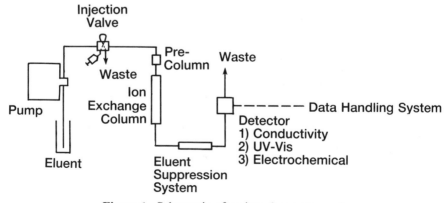

Figure 1. Schematic of an ion chromatograph.

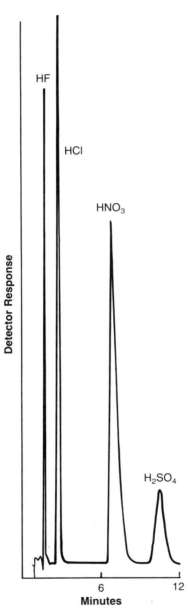

Figure 2. Determination of HF, HCl, HNO$_3$, and H$_2$SO$_4$.

3. Determination of acidic thermal decomposition products of polymer materials.
4. Determination of nutrients and metabolic by-products in fermentation broths—amino acids, carboxylic acids, carbohydrates, etc.

5. Determination of chlorine, bromine, iodine, phosphorus, and sulfur in organic molecules after an appropriate oxidation.
6. Determination of fluoride, chloride, and sulfate in phosphoric acid. See Figure 3.

Limitations

Ion chromatography is limited by the inability to vary, in most cases, the ionic species column selectivity, making it virtually impossible with present

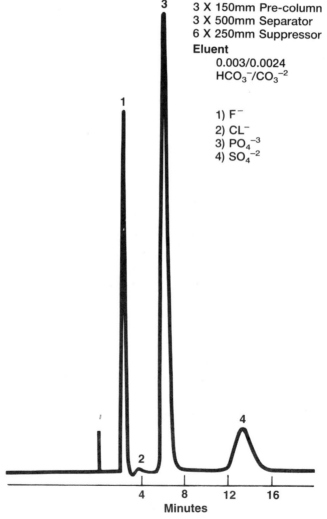

3 X 150mm Pre-column
3 X 500mm Separator
6 X 250mm Suppressor
Eluent
        0.003/0.0024
        HCO_3^-/CO_3^{-2}

1) F^-
2) CL^-
3) PO_4^{-3}
4) SO_4^{-2}

Figure 3. Determination of fluoride, chloride, and sulfate in phosphoric acid.

technology to control the elution order of the various ionic species in the sample. Another major limitation is the difficulty of analyzing a trace ionic species in the presence of a preponderance of another ionic species.

References

(1) Gjerde, D. T.; Fritz, J., Ion Chromatography; 2nd edition, Hüthig, Verlag: Heidelberg, 1987; 283.
(2) Smith, F.; Chang, R., The Practice of Ion Chromatography; John Wiley and Sons: New York, 1982; 218.
(3) Weiss, J., Handbook of Ion Chromatography; Dionex Corp.: Sunnyvale, CA, 1986; 244.

GEL PERMEATION CHROMATOGRAPHY

Use

Gel permeation chromatography separates molecular species by size. Its primary application is to define the molecular weights and molecular weight distribution of polymeric materials. It also finds utility in separation of additives in polymeric resins and of small molecule mixtures.

Sample

Approximately 10 mg quantities of either a solid or a liquid are required for evaluation. The sample must be miscible with a solvent which is compatible with the separating medium and have a refractive index which differs from the solvent by approximately 0.01 unit or more. Smaller quantities of samples may be evaluated when solute-solvent refractive index differences are large. All types of materials capable of dissolution at temperatures of 145°C or less can be evaluated.

Normal sample dissolution, instrument operation and calculation times vary from 0.5 to 2 hours and depend primarily on the nature of the sample.

Principle

Size-separation of molecular species is accomplished by eluting the solution through columns which contain crosslinked polystyrene (gel) or silica of varying porosity. Separation of molecular species depends on the degree of permeation into and through the medium. Molecules larger than the maximum pore size of the medium pass through the column's interstitial volume. Molecules smaller than the maximum pore size enter the medium and are size separated. Retention times for larger molecules are shorter than for smaller molecules. Weight concentration of eluting molecular species is automatically measured by a continuous recording of the differential refractive index of solution and solvent. Alternately, the UV or IR absorption, the scattered light, or the viscosity of the eluting solution can be measured. Molecular parameters are obtained by calibrating the columns with well-characterized fractions. It is advantageous to use fractions having narrow molecular weight distributions which cover a wide molecular weight range.

Comparisons of chromatograms obtained on the same types of materials provide qualitative evaluation of similarities or differences in molecular parameters. Calibration of the medium used to separate molecular species allows calculation of weight and number average molecular weights, and molecular weight distribution. Calibration will also allow estimates of small molecule molecular weights. In the absence of adsorption and/or excessive dispersion, weight average molecular weights compare favorably with similar values obtained by classical techniques. Number average molecular weights calculated from gel permeation chromatograms are, in general, lower than those obtained by classical techniques. Low molecular weight species are more strongly affected by adsorption and axial diffusion phenomena than are high molecular weight species. Use of the proper solvent may reduce adsorption effects. Dispersion effects can be reduced by reducing solvent flow rate.

A block diagram of a gel permeation chromatograph is given in Figure 1.

Applications

1. Gel permeation chromatography is used to find and compare the molecular weight distribution of polymer samples. This is demonstrated in Figure 2 where the chromatograms of two different styrene vinylbenzyl chloride copolymers are shown. The calibration of the columns shows that the sample with the higher molecular weight has the greater polydispersity.

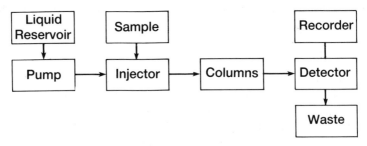

Figure 1. Block diagram of a gel permeation chromatograph.

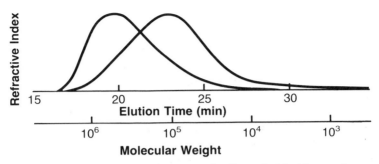

Figure 2. Chromatograms of styrene vinylbenzyl chloride copolymers.

2. Branching in polymers can be observed by comparing the elution time to the reduced viscosity of the fraction. Branching affects each parameter in a different way so combining the information gives a measure of the degree of branching in each part of the distribution.
3. Size distributions of colloidal particles can also be measured by gel permeation chromatography.
4. Different species in a mixture can be isolated and quantified. This is shown in Figure 3 for a neopentyl glycol/isophthalate condensate. Three peaks are readily evident in this chromatogram. The peak at the lowest retention volume (highest molecular weight) results primarily from 4-4 and 4-3 alcohol/acid combinations. The intermediate peak results from 2-2 and 2-1 alcohol/acid combinations. The lowest molecular weight peak results from the unreacted neopentyl glycol.
5. Oligomers can be isolated from the sample through gel permeation chromatography isolation.

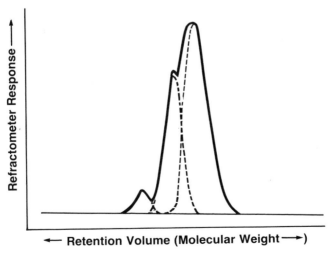

Figure 3. Chromatogram of neopentyl glycol/isophthalate condensate.

Limitations

Applicability of this technique assumes the existence of a solvent compatible with the sample, the separation media, and the equipment. GPC is most often used when molecular weight distributions are sought. The technique, however, separates on the basis of particle size, not weight. The distinction becomes critical when polymers exist in multiple configurations where sizes are equal, but the weights are not. A common frustration when interpreting results occurs from difficulty in relating the molecular weight of a sample to those for calibration standards, especially when there may be branching. GPC does not separate polymers larger than the largest pores in the separation media, and shearing within the columns may cause polymer degradation to even smaller polymers. The technique is usually not good for polymers in excess of ten million molecular weight.

References

(1) Kremmer, T.; Boross, L., Gel Chromatography; John Wiley and Sons: New York, 1979.

(2) Provder, T. ed.; Size Exclusion Chromatography; American Chemical Society: Washington, D.C., 1984.

(3) Yau, W. W.; Kirkland, J. J.; Bly, D. D., Modern Size Exclusion Chromatography; John Wiley and Sons: New York, 1979.

CHAPTER 5

Chemical Analysis, Electrochemistry and Atomic Spectroscopy

Daniel E. Bause, Richard J. Williams, Kurt Theurer

CLASSICAL CHEMICAL ANALYSIS

Use

Almost any element or functional group can be determined by a purely chemical approach, tailored for the particular application. This is the time-honored "classical" approach, and the literature on this subject is voluminous. While instrumental methods, because of their rapidity, now handle many determinations, there still remain those for which a "classical" procedure is the preferred method, especially when absolute accuracy and high precision are required, or when standards must be analyzed prior to calibration of some other technique. The major disadvantage is the time-consuming nature of many of the procedures required to isolate the desired constituent in a unique form suitable for determination.

Sample

Any material—gas, liquid or solid—can be analyzed by chemical techniques. Sample quantities are dependent upon the element or group sought and upon the anticipated concentration levels. An adequate sample is usually 1–5 grams, but successful analyses can often be made with a few milligrams; on the other hand, in the case of trace determinations, a few hundred grams of sample may be required.

Principle

There is but one principle which forms the basis for all classical techniques: stoichiometry. The chemical reaction between two or more species in solution produces a compound of precisely known composition, which is then accurately measured by one of the following: gravimetry, volumetry, photometry, or turbidimetry.

Applications

Gravimetry is the technique commonly used for macro determinations of such constituents as silica and sulfate. Following complex separation schemes, it is also employed for the determination of tungsten, tantalum, niobium, gallium, and many of the rare earths.

Volumetry, probably the most frequently used end point in chemical analysis schemes, can be applied to both aqueous and non-aqueous media, and is applicable to both macro and micro concentration levels. Typical routine procedures include acidimetry and alkalimetry, argentimetry (halides, cyanides), iodimetry and iodometry (sulfite, copper, antimony), oxidation-reduction titrimetry (iron, titanium, uranium), compleximetric titrimetry (calcium, magnesium, nickel complexed with ethylenediamine tetraacetic acid) and non-aqueous titrimetry (polymer end groups, saponification number, hydroxyl). End point detection is either visual with a suitable colorimetric indicator or potentiometric when solutions are highly colored or an increase in sensitivity is required.

Photometry is now usually limited to trace level determinations of anionic species, because atomic absorption spectrophotometry has become the method of choice for trace metal analysis. Some of the more common photometric procedures include analysis for chlorine, cyanides and phenols in wastewaters; ammonia by Nesslerization; nitrates and nitrites; and phosphorus—macro as well as trace.

Turbidimetry, a special case of photometry, is employed when the analyte species reduces the light transmission as a result of the opacity of uniform, suspended particles rather than by light absorption. Turbidimetry is commonly used as an end point to determine small amounts of suspended silver chloride (trace chlorine analysis) and of barium sulfate (trace sulfur and sulfate analysis). When the suspended particles are so small that cloudiness is not visible, but the particles can still reduce light transmission by scattering, the measurement is called nephelometry.

Limitations

The classical techniques usually require more sample manipulation than most other analytical methods. Thus, their major disadvantage is their time-consuming nature. Additionally, unwanted precipitates or colored species may also be present as interferences.

References

(1) Kolthoff, I. M.; Sandell, E. B.; Meehan, E. J.; Bruckenstein, S., Quantitative Chemical Analysis; 4th ed., Macmillan: London, 1969.

(2) Meites, L. ed., Handbook of Analytical Chemistry; McGraw-Hill: New York, 1963.

(3) Skoog, D. A.; West, D. M., Fundamentals of Analytical Chemistry; 3rd ed., Holt, Rinehart and Winston: New York, 1976; Chapters 5–16.

FUNCTIONAL GROUP ANALYSIS

Use

Quantitative functional group methods are used in organic analysis to establish compound purity, determine analyte concentration in complex mixtures and help in the identification of unknown compounds by establishing the equivalent weight. Functional group analysis by chemical methods has the advantage that reference materials and calibrations are not required.

Sample

Methods exist or can be developed to handle materials in any physical form. Usually a few milligrams to 1 gram of material is adequate for most analyses.

Principle

The chemical method is based on the specific reaction between the functional group and a specific reagent, followed by a quantitative determination of either the reaction product or excess reagent. Reactions that consume or liberate acid, base, oxidant, reductant or a gas are particularly useful, because quantitative measurements can be readily made with such substances.

Applications

Applications using functional group analysis are varied and numerous. A few typical examples are:

1. Amine and carboxyl end groups in nylon.

$$RNH_2 + HX \rightarrow RNH_3^+X^-$$
$$RCOOH + NaOH \rightarrow RCOO^-Na^+ + H_2O$$

2. Carboxyl end groups in poly(ethylene terephthalate).

$$RCOOH + NaOH \rightarrow RCOO^-Na^+ + H_2O$$

3. Hydroxyl groups in high molecular weight alcohols and complex mixtures.

$$
ROH + CH_3-\overset{\displaystyle O}{\overset{\|}{C}}\rightarrow CH_3-\overset{\displaystyle O}{\overset{\|}{C}}-OR + CH_3COOH
$$

$$
\begin{array}{c}
\diagdown O \diagup \\
CH_3 - C \\
\| \\
O
\end{array}
$$

4. Determination of carbonyl compounds.

$$
R-\overset{\displaystyle O}{\overset{\|}{C}}-R' + NH_2OH \cdot HCl \rightarrow R-\overset{\displaystyle NOH}{\overset{\|}{C}}-R' + H_2O + HCl
$$

5. Olefinic unsaturation in fatty acids, vinyl esters and other unsaturated compounds.

$$
\overset{\diagdown}{\underset{\diagup}{C}}=\overset{\diagup}{\underset{\diagdown}{C}} + ICl \rightarrow -\overset{|}{\underset{|}{C}}-\overset{|}{\underset{|}{C}}- \\
 I \;\; Cl
$$

Limitations

One of the main problems in functional group analysis is the dissolution of the sample in the appropriate solvent. Another problem is that many reactions tend to reach an equilibrium and a means must be found to drive the reaction to completion.

References

(1) Cheronis, N. D.; Ma, T. S., Organic Functional Group Analysis by Micro and Semimicro Methods; John Wiley and Sons: New York, 1964; 696.

(2) Connors, K. A., Reaction Mechanisms in Organic Analytical Chemistry; John Wiley and Sons: New York, 1973; 634.

(3) Siggia, S., Instrumental Methods of Organic Functional Group Analysis; Wiley-Interscience: 1972; 428.

COMBUSTION ANALYSIS

Use

Combustion analysis is used to determine C, H, N, O, S, P, and halogens in a variety of organic and inorganic materials at the trace to percent level.

Sample

Any gas, liquid or solid can be analyzed. Sample quantities are dependent upon the concentration level of the analyte. A few milligrams of material are often sufficient for a precise assay.

Principle

Combustions are performed under controlled conditions in the presence of catalysts. Oxidative combustions are most common. The element of interest is converted into a reaction product which is readily determined by a variety of techniques such as gas chromatography, ion-selective electrode, titrimetry, ion chromatography, and colorimetric measurement. Combustion techniques commonly used in the analytical laboratory are as follows:

Automated C, H, N, and O Analysis

Simultaneous analysis of C, H, and N in organic materials is accomplished by combustion in an oxygen atmosphere at temperatures up to 1000°C. The gaseous combustion products are flushed with helium through a reductant so that the emerging helium carrier gas contains only CO_2, H_2O, and N_2. Quantification is obtained by gas chromatography. For analysis of O, the combustion is performed in a helium atmosphere over platinized carbon. The CO formed is converted to CO_2 by passage over CuO, and measured in the same manner as for analysis of C. A 2–3 mg sample of liquid or solid is required for a single C, H, N, or O determination.

Schöniger Oxygen Flask Combustion

In this method, an organic sample is ignited in a thick-walled erlenmeyer-type flask equipped with a solid glass stopper. Attached to the stopper is a platinum basket in which the sample is ignited. The gases formed are immediately absorbed by a layer of appropriate absorbing liquid in the flask, resulting in inorganic ions (F^-, Cl^-, Br^-, I^-, SO_4^{-2}), which are then quantitatively determined by the best appropriate method. Sample weight is limited to less than 200 mg.

Parr Oxygen Bomb

This technique is similar to the Schöniger oxygen flask combustion, except that up to 1 gram of sample can be ignited at 40 atmospheres of oxygen in a stainless steel vessel. The technique has better sensitivity (detection limits of 20 ppm Cl and S) and is more suitable for volatile samples.

Combustion Techniques for Total Organic Carbon

A few milliliters of sample are added to an ampoule containing 25% phosphoric acid and potassium persulfate. The ampoule is flame sealed and placed in an autoclave for one hour. The persulfate oxidizes any organic carbon to CO_2. The CO_2 is measured by IR after opening the ampoule inside the instrument. The method has a detection limit of 1 ppm carbon. Inorganic bicarbonate/carbonate can be measured by omitting the persulfate.

Applications

1. Determination of organic-bound halogens in a variety of organic materials such as epoxy molding resins, halogenated hydrocarbons, brominated resins, etc.
2. Phosphorus content in flame retardant materials.
3. Monitoring total organic carbon in water samples such as sewage or industrial wastewater.
4. Determination of total sulfur in petroleum products.
5. Identification of organic compounds based on precise determination of elemental composition.

Limitations

Combustion techniques are limited by the inability to always obtain complete combustion especially when dealing with polymeric and highly fluorinated materials and the high blanks for chlorine and to a lesser amount sulfur when using either Schöniger or Parr Oxygen Bomb combustions. A limitation of organic elemental analysis is the inability of obtaining a homogeneous sampling of many industrial materials when only 1–2 mg is weighed out.

References

(1) Ingram, G., Methods of Organic Elemental Analysis; Reinhold: New York, 1962.

(2) Kirsten, W. J., Organic Elemental Analysis Ultramicro, Micro and Trace Metals; Academic Press: New York, 1983; 146.

(3) Ma, T. S.; Ritter, C. R., Modern Organic Elemental Analysis; Marcel Dekker: New York, 1979; 581.

ION-SELECTIVE ELECTRODE ANALYSIS

Use

Ion-selective, or specific ion electrodes, are used for the measurement of concentration of ionic species or dissolved gases in solution. Techniques most commonly used are direct reading, with or without standard additions and titrations with the electrode acting as the end-point detector.

Sample

Samples may be aqueous solutions, or solids which can be rendered soluble in water. In certain cases, measurements may also be performed in selected non-aqueous solvents such as methanol, acetone, or dioxane. The quantity of sample required depends upon the level of analyte sought; generally determinations are made in volumes of 10 to 100 ml.

Principle

The detection of an ionic species depends upon the potential developed across a conducting membrane separating the reference solution from the sample. This membrane system is so designed that the only conducting species within it is a specific ion (analyte). The developed potential is then proportional to the level of free ions in the sample solution. From the Nernst equation,

$$E = E_0 \pm m \, \text{Log} \, A,$$

where E is the potential, E_0 is the reference potential (a constant) and m is the electrode slope, the ionic activity, A, of the sample may be determined. In the case of gas-sensing electrodes, the separating membrane is made to be permeable to the analyte (NH_3, CO_2, etc.) but impermeable to water and other ionic species, thus providing the selectivity needed for measurement.

Applications

Some typical applications include: fluoride and nitrate determination in stainless steel pickling baths, ammonia in air and stack gases, cadmium in plating baths, calcium in blood, chloride in boiler feed water, cyanide in gold and cadmium plating baths, alkali metals (Na^+, K^+) in biological fluids. Figure 1 shows a typical calibration curve for fluoride ion. Typical detection limits by ion selective electrodes are given in the following table.

Electrode	Limit of Detection M (ppm)
F^-	10^{-7}M (0.002)
Cl^-	10^{-5}M (0.35)
Br^-	10^{-6}M (0.08)
I^-	10^{-8}M (0.001)
SCN^-	5×10^{-6}M (0.3)
CN^-	10^{-6}M (0.03)
S^{-2}	10^{-7}M (0.003)
NO_3^-	5×10^{-6}M (0.31)
ClO_4^-	10^{-6}M (0.1)
BF_4^-	2×10^{-6}M (0.1)
Ag^+	10^{-7}M (0.01)
Pb^{+2}	10^{-7}M (0.02)
Cd^{+2}	10^{-7}M (0.01)
Cu^{+2}	10^{-9}M (0.0001)
Hg^{+2}	10^{-8}M (0.002)
Ca^{+2}	10^{-6}M (0.02)
K^+	10^{-6}M (0.04)
NH_3	10^{-7}M (0.002)
CO_2	2×10^{-6}M (0.1)

Limitations

Ion-selective electrodes measure the activity, rather than the concentration of ions in solutions, although with proper calibration a concentration value can be obtained. Because of the electrode logarithmic response, the accuracy and precision of the determinations are sometimes poorer than with other techniques.

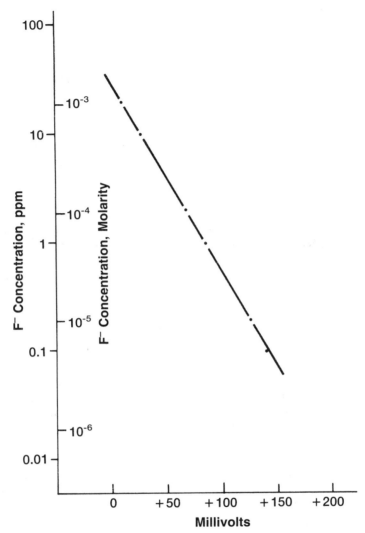

Figure 1. Typical calibration curve for the measurement of fluoride ion.

References

(1) Bailey, P. L., Analysis with Ion-Selective Electrodes; Heyden and Sons; London, 1976; 228.

(2) Baiulescu, G. E.; Cosofret, V. V., Applications of Ion-Selective Membrane Electrodes in Organic Analysis; John Wiley and Sons: New York, 1977; 235.

(3) Mooly, G. J.; Thomes, J. D. Ion-Selective Electrode, Rev. 6, 1984, 209.

RADIOACTIVE TRACER ANALYSIS

Use

Radioactive tracers are used to follow the behavior of atoms or groups of atoms in a chemical reaction or physical transformation. A radioactive tracer unequivocally labels the particular atoms to be traced, regardless of what may happen to them in a complicated system. The specificity and sensitivity of detection of radioactive isotopes is extremely good. Less than 10^{-8} grams can often be detected.

Sample

Radioactive materials to be counted can be solid, liquid, or gas. Sample preparation is normally kept to a minimum. For liquid scintillation counting, the radioactive substance is put into intimate contact with the scintillation medium by dissolving it, suspending it, or immersing it in the liquid solution of fluors.

Principle

Radioactivity is the phenomenon of the spontaneous disintegration of metastable atomic nuclei with the emission of energetic radiations. There are three principle types of radiation emitted by radioactive substances, designated as alpha, beta, and gamma rays. They differ in their physical nature, and from the point of view of their detection and measurement they are distinguished principally by the manner in which they interact with matter to lose energy and produce ionization. Liquid scintillation counting is commonly used for counting a wide range of alpha and beta emitting radioisotopes in many chemical forms. The radioactivity is detected by means of a solution of fluors and a photo-multiplier tube. The scintillation solution converts the energy of the primary particle emitted by the radioactive sample to visible light and the phototube responds to this light energy by producing a charge pulse which can be amplified and counted by a scaling circuit.

Applications

Applications using radioactive tracers are varied and numerous. A few typical examples are:

1. Study of inorganic, organic and polymer synthesis mechanisms.

2. Copolymer composition by infrared spectroscopy using ^{14}C tagged polymers to calibrate spectra.
3. Catalyst poisoning including decomposition and interactions.
4. Measurement of homogeneous and heterogeneous reaction rates.
5. Biochemical metabolism studies.
6. Chemical behavior of tagged polymeric materials tailored for specific end uses.
7. Solubility, co-precipitation and separation studies.
8. Measurement of nutrient uptake in vegetation.
9. Metal-metal self-diffusion studies at elevated temperatures.
10. Measurement of permeation rates through thin films.
11. Corrosion studies of surfaces.
12. Stability of organic solvents and additives over long time storage.

Limitations

Definite limitations of the radioactive tracer method are the absence of known radioactive isotopes of suitable half-life for a few elements, especially oxygen and nitrogen and the availability of radiochemically pure compounds. When following a tagged compound through an analytical scheme or chemical process, it is essential that the compound be tagged with an atom that is not readily exchangable with similar atoms in other compounds. If the tagged atom is part of a multiple decay scheme, it may be necessary to separate it from its radioactive daughters before counting. Liquid scintillation counting is subject to a variable counting efficiency caused by quenching of light by certain chemical impurities; other impurities may introduce chemiluminescence which give an unknown and variable background. The US Nuclear Regulatory Commission may require the user to obtain a materials license before engaging in extensive radiochemical operations.

References

(1) Barnes, W. E. ed., Basic Physics of Radiotracers; Vol. 1 and 2, CRC Press: Boca Raton, FL, 1983; pp 166, 205.

(2) Kyrs, M.; Prikrylova, K., J. Radioanal. Nuc Chem., 81(2), 1984, pp 227–233.

(3) Mundy, J. N.; Rothman, S. R.; Fluss, S. J.; Smedskjaer, L. C., eds., Methods of Experimental Physics, 21, Solid State: Nuclear Methods; Academic Press: Orlando, FL, 1983, 504.

(4) Yoshioka, H.; Kambara, T., Talanta, 31(7), 1984, pp 509–513.

POLAROGRAPHY AND VOLTAMMETRY

Use

Polarography and voltammetry are electrochemical techniques which provide quantitative and qualitative information about redox active (capable of oxidation or reduction) components. In certain cases the technique can be extremely sensitive to trace levels of analyte.

Sample

Solids, liquids or gases can be analyzed. As little as ten micrograms of a polarographically active material will give a measurable signal.

Principle

Polarography and voltammetry are based on the principle that redox active materials can accept or lose electrons at a given potential with respect to a reference electrode. The working electrode, where the electrochemical reaction occurs, is usually the dropping mercury electrode (polarography), but other working electrodes (voltammetry) such as platinum, carbon, and hanging mercury drop can be used to advantage. The magnitude of current flow at the potential where the analyte being studied oxidizes or reduces is proportional to the concentration of the analyte, thus allowing a quantitative determination. The sample being analyzed is first dissolved and then mixed with a supporting electrolyte that provides conductivity and the best environment for the analysis. An appropriate technique such as differential pulse polarography (DPP) or stripping voltammetry is then used to analyze the sample. A relatively new technique, square wave voltammetry, has replaced differential pulse polarography for some applications. In the square wave mode, the scan rate is significantly faster and sensitivity is enhanced.

Qualitative information about redox couples can be supplied by operating in the cyclic voltammetry mode. This method enables a wide potential range to be rapidly scanned for reducible or oxidizable species. A schematic of a polarograph is given in Figure 1.

Applications

1. METALS—Almost thirty metals can be analyzed at levels as low as 50 ppb by DPP at a dropping mercury electrode. With the proper choice of supporting electrolyte, the oxidation state of the metal can be determined in many cases. Stripping voltammetry using the differential pulse mode

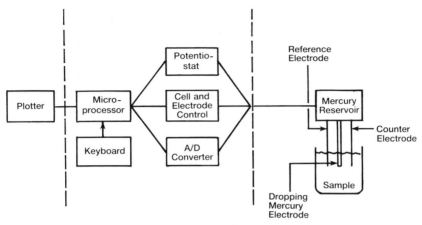

Figure 1. Schematic of a polarograph.

can be used with advantage for ultra trace level work. For example, Pb, Cd, Zn and Cu can be determined at the 1 ppb level in tap water using stripping techniques. Figure 2 shows a differential pulse anodic stripping voltammogram of tap water containing Zn, Cd, Pb, and Cu at ppb levels.

2. INORGANIC COMPOUNDS AND IONS—There are a wide range of inorganic compounds which can be analyzed by polarography and vol-

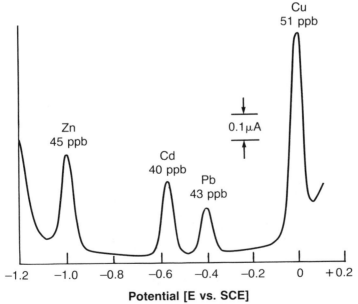

Figure 2. Differential pulse anodic stripping voltammetry in 0.2M ammonium citrate, pH 3.

tammetry. Many nitrogen-containing compounds such as NH_3, NH_2OH, NH_2NH_2 and N_3^- can be analyzed using DPP to sub-ppm levels. Often derivatization can lead to improved detection levels. Many common anions are polarographically active and can be detected at sub-ppm levels, e.g., SO_3^{-2}, $S_2O_3^{-2}$, BrO_3^-, Br^-, and CN^-. Cathode stripping techniques afford ppb sensitivity for Cl^-, I^-, SeO_3^{-2}, and S^{-2}.

3. ORGANIC COMPOUNDS—The ability to analyze organic compounds by polarography is determined by the presence or absence of electroactive functional groups. Electroactive functional groups include aldehydes, ketones, peroxides, oximes, nitro compounds, amines, and thiocyanates. Formaldehyde can be readily determined at ppb levels by DPP. Figure 3 shows a polarogram of a mixture of maleic and fumaric acids at the ppb level. It is also possible to convert inactive species to ones which are amenable to polarography by the formation of electroactive derivatives. For example, aminocaproic acid in nylon can be determined after the addition of formaldehyde and quantitative formation of the Schiff base.

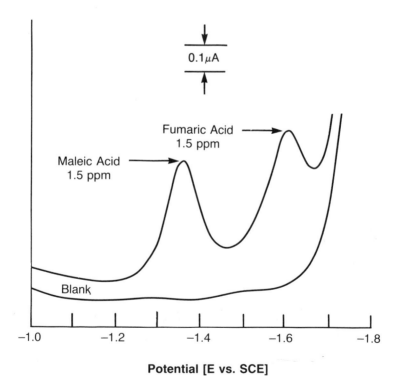

Figure 3. Differential pulse polarography of maleic and fumaric acids in 0.2M phosphate buffer, 1M NH_4Cl, pH adjusted to 8.2.

Limitations

The most serious limitation to the dropping mercury electrode is the ease with which it is oxidized. The dropping mercury electrode can be employed only for the analysis of reducible or very easily oxidizable substances. Care must also be taken with the apparatus used to generate the mercury drop, otherwise erratic and nonreproducible drop times and sizes will occur. Anodic stripping methods, which are very sensitive, are limited to about 15 elements.

References

(1) Bard, A.; Faulkner, L., Electrochemical Methods; Wiley: New York, 1980.

(2) Bond, A. M., Modern Polarographic Methods in Analytical Chemistry; Marcel Dekker: New York, 1980.

(3) Kissinger, P. T.; Heineman, W. R. eds., Laboratory Techniques in Electroanalytical Chemistry; Marcel Dekker: New York, 1984.

(4) Skoog, D. A.; West, D. M., Principles of Instrumental Analysis; Holt, Rinehart and Winston: New York, 1971, pp 553–587.

ISOTACHOPHORESIS ANALYSIS

Use

Isotachophoresis analysis can be used to determine both low and high molecular weight charged substances with a minimum of sample pretreatment.

Sample

The sample must be liquid (miscible with water) or a water soluble solid. Samples as little as 1 to 10 microliters can be analyzed. Sensitivities of many components are sufficient to allow detection at the picomole level.

Principle

Isotachophoresis involves a migration of ionic species in an electric field. Isotachophoresis will take place when an electric field is applied to a system of electrolytes (leading and terminal electrolyte) in a small Teflon capillary. A sample solution is introduced between the leading and terminal electrolytes. The leading electrolyte must contain only one ionic species, the leading ion, having the same sign as the sample ions to be separated, and an effective mobility higher than that of any of the sample ions. The terminal electrolyte must contain only one ionic species, the terminal ion, having the same sign as the sample ions to be separated, and an effective mobility lower than that of any of the sample ions. Separation is controlled by choice of leading and terminal electrolyte. Four different types of detection systems are used in isotachophoretic analysis: UV-absorbance, conductometric, potential gradient, and thermometric detection. A schematic of an isotachophoresis instrument is given in Figure 1.

Applications

1. Any charged species, low or high molecular weight.
2. Any inorganic ion can be analyzed—anion, cation, and metal complexes.
3. As for organic substances, those having the following functional group can be analyzed.
   A. COOH—Carboxylic acids.
   B. SO_3H—Sulfonic acids/surfactants.
   C. NH_2—Amines.
   D. $^+N(R)_4$—Quaternary ammonium compounds.
   E. $H_2N-CH-COOH$—Amino acids

             |

           R

   F. ⬡—OH—Phenol and substituted phenols.

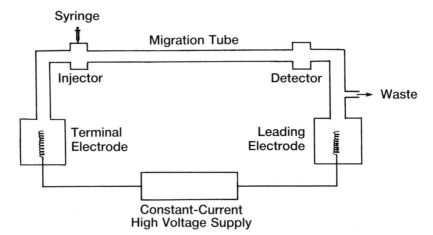

Figure 1. Schematic of an isotachophoresis instrument.

Limitations

The main limitations of analytical isotachophoresis are the difficulties of isolating unknowns for identification and the time required to choose the appropriate leading and terminal electrolytes.

References

(1) Everaerts, F. M. ed., Analytical Isotachophoresis; Elsevier: 1981, 234.

(2) Everaerts, F. M.; Beckers, J. L.; Verheggen, T., Isotachophoresis—Theory, Instrumentation and Applications; Elsevier: 1976, 418.

(3) Everaerts, F. M.; Verheggen, T.: Reijenga, J. C., Trends in Analytical Chemistry, 2, 1983, 188.

ATOMIC ABSORPTION SPECTROSCOPY

Use

Atomic absorption spectroscopy is a technique for determining the concentration of metallic elements in solution. The solvent most commonly used is water, although organic solvents can be effectively handled in many instances. Atomic absorption is specific and characteristic for each element. A limitation of the method is that only one element may be determined at a time. When lower detection limits are required, atomic absorption can serve as a method which is complementary to ICP spectrometry.

Sample

The sample may be either liquid or solid, but must be diluted or dissolved in a suitable medium. The quantity required may vary and is dependent upon the element(s) sought and the anticipated concentration level. Extremely low concentrations may be determined by using concentrative techniques such as evaporation or solvent extraction.

Principle

An emission source, commonly a hollow cathode lamp or an electrodeless discharge lamp, produces the line spectrum of a particular element. When a specific wavelength of this radiation is passed through a vapor containing ground state atoms of the same element, some of this radiation is absorbed. The decrease in the signal observed by the photomultiplier tube is a function of the quantity of the element in the vapor.

There are several common methods for producing the sample vapor: aspiration of the sample solution into a flame, which may be air-acetylene or nitrous oxide-acetylene; atomization in a graphite furnace (commonly called flameless); by converting the metallic element to its hydride, or by cold vapor generation for mercury. In Figure 1 is a schematic of an atomic absorption spectrophotometer.

Applications

Atomic absorption procedures can be applied to almost all types of samples including alloys, polymers, ceramics, and wastewaters. The detection limits for flameless AA are usually one to two orders of magnitude lower than the corresponding flame AA method. The determination of alkali metals is handled better by AA methods than ICP spectroscopy. In Figure 2 is given a typical calibration curve for iron using an air-acetylene flame. Typical detection limits by atomic absorption are given in the following table:

	Flame (mg/L)	Flameless (mg/L)
Aluminum	0.1	0.003
Antimony	0.2	0.003
Arsenic	0.002(1)	0.001
Barium	0.1	0.002
Beryllium	0.005	0.0002
Cadmium	0.005	0.0001
Chromium	0.05	0.001
Copper	0.02	0.001
Gold	0.1	0.001
Iron	0.03	0.001
Lead	0.1	0.001
Manganese	0.01	0.0002
Mercury	—	0.0002(2)
Nickel	0.04	0.001
Potassium	0.01	—
Silicon	0.02	0.001
Silver	0.01	0.0002
Sodium	0.002	—
Tin	0.8	0.005
Vanadium	0.2	0.004
Zinc	0.005	0.00005

(1) by gaseous hydride method
(2) by cold vapor technique

Limitations

As with other analytical methods that deal mainly with solutions, atomic absorption spectroscopy is dependent upon the ability to get the sample into

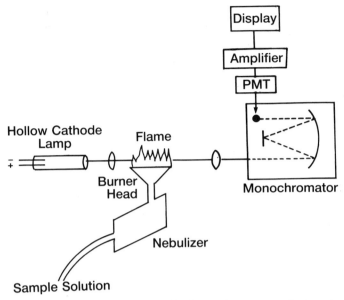

Figure 1. Schematic of an atomic absorption spectrophotometer.

solution. The calibration range of AAS is also rather small and is usually limited to one or two orders of magnitude. This narrow calibration range means frequent dilutions are necessary for more concentrated solutions. Unlike multielement ICP spectroscopy, AAS is limited to the determination of one element at a time. Chemical interferences are also encountered in

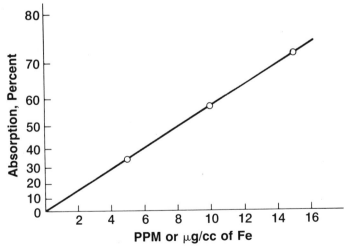

Figure 2. Atomic absorption spectroscopy calibration curve for iron.

AAS. Anything that prevents or suppresses the formation of ground state atoms in the flame is considered a chemical interference. The addition of a releasing agent or switching to a nitrous oxide-acetylene flame may be successful in overcoming chemical interferences.

References

(1) Cantle, J. E., ed., Atomic Absorption Spectrometry; Elsevier: New York, 1982.

(2) Price, W. J., Analytical Atomic Absorption Spectrometry; Heydon: London, 1972.

(3) Slavin, W., Atomic Absorption Spectroscopy; Wiley-Interscience: New York, 1968.

EMISSION SPECTROGRAPHIC ANALYSIS

Use

Emission spectrographic analysis is a technique for simultaneously determining the concentration of about 70 elements in inorganic and organic matter.

Sample

Usually 20 milligrams of an inorganic solid sample is adequate for semi-quantitative determinations. As little as one milligram will usually supply much information on sample composition. Considerably larger amounts of sample are desirable for organic substances and for samples with a volatile matrix.

Principle

When a substance is correctly excited by an electrical arc or spark, every element present emits light at wavelengths that are specific for that element. The light emitted passes through a narrow vertical slit and is dispersed with a grating or prism. The spectral lines produced are generally recorded on a photographic plate or film. The location and intensity of the lines produced by the sample are compared, either visually or with a photo-electric densitometer, with the lines produced by suitable standards of known composition (Figure 1).

A schematic of an arc/spark emission spectrograph is given in Figure 2. An important variation of this technique is presented by an instrument

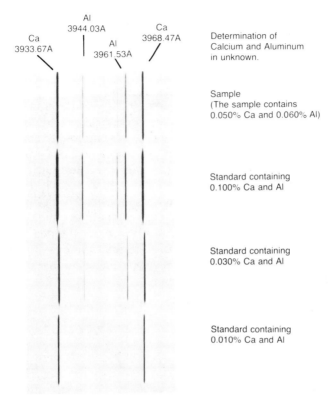

Figure 1. Emission spectrographic analysis of Ca and Al in polymer ash.

called a direct-reading spectrograph. In this case, the spectral line intensities are recorded directly by photomultiplier tubes located at specific, predetermined wavelengths, one photomultiplier tube being used for each element. Through the use of calibration standards, the observed line intensities may be directly related to concentrations. In contrast to a spectrograph employing photographic plates, which is limited to trace analyses (0.001–1%), the

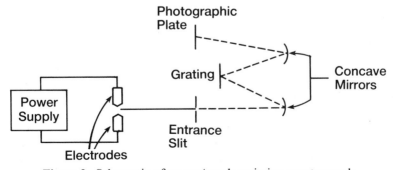

Figure 2. Schematic of an arc/spark emission spectrograph.

direct-reading spectrograph may be designed for both trace and major element concentrations in specific matrices.

Applications

Emission spectrographic techniques can be applied to almost every type of sample. Areas of investigation include chemicals, ceramics, metals and alloys, plastics, agricultural products, foodstuffs and water analyses. For volatile samples such as organic solvents, mineral acids, and wastewater, preconcentration of the sample makes it possible to determine most elements at the low parts-per-billion level.

The direct-reading spectrograph, on the other hand, is designed to handle a very limited range of materials, such as metal alloys, where matrix variability is slight. Calibration standards must be generated for each significantly different alloy. Samples must be prepared in the form of a 1″ diameter button with a highly polished flat surface; approximate sample size required is 40–50 grams.

Typical detection limits are tabulated below:

Element	Micrograms	Element	Micrograms
Aluminum	0.05	Manganese	0.05
Barium	0.5	Molybdenum	0.05
Beryllium	0.002	Nickel	0.1
Boron	0.1	Palladium	0.02
Calcium	0.01	Silicon	0.2
Chromium	0.05	Silver	0.01
Cobalt	0.05	Tin	0.1
Copper	0.02	Titanium	0.02
Gold	0.2	Vanadium	0.05
Iron	0.1	Zirconium	0.05
Lead	0.2		

Limitations

Emission spectrographic analysis, like any other emission technique, is hindered by spectral interference problems. Sometimes an alternate line is available, but often this secondary line does not have the sensitivity required for the analysis. Another drawback of the arc emission method is the time-consuming, labor intensive nature of the procedure. The electrodes are prepared, analyzed, and, if photographic detection is being used, the plate must be developed and the emission lines identified and quantified. Concentrations are semiquantitative at best for the photographic plate detection. The major drawback of the direct-reading spectrograph is the availability of suitable solid standard materials. Therefore, this method is usually reserved for metal alloys.

References

(1) Kolthoff, I. M.; Elving, P. J. eds., Treatise on Analytical Chemistry; Part I, Vol. 6, Interscience: New York, 1965, Chapter 64.

(2) Sawyer, R. A., Experimental Spectroscopy; 3rd ed., Dover: New York, 1963.

(3) Willard, H. H.; Merritt, Jr., L. L.; Dean, J. A., Instrumental Methods of Analysis; 4th ed., Van Nostrand: Princeton, NJ, 1965, Chapter 10.

FLAME EMISSION SPECTROMETRY

Use

Flame emission spectrometry is a simple and convenient method for analyzing solutions of organic and inorganic materials for individual metals. This method is particularly suited to alkali and alkaline earth metals.

Sample

1–5 grams of a dissolved sample, either solid or liquid, is adequate for analyses. Extremely low concentrations of metals can often be determined by suitable concentration procedures.

Principle

A solution of the sample to be analyzed is aspirated into a flame possessing the thermal energy required to excite the element to a level at which it will radiate its characteristic bright line emission spectrum. The intensity and frequency of the radiation is measured photoelectrically using a spectrometer. The specific frequency of the radiation identifies the element producing it, while the intensity of the radiation at the specific frequency is proportional to the amount of the element present, affording a quantitative analysis.

A diagram of a typical flame emission spectrometer is given in Figure 1.

Applications

Flame emission spectrometric techniques, both qualitative and quantitative, can be applied to clinical materials (serum, plasma, and biological

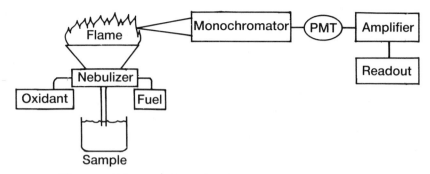

Figure 1. Diagram of a typical flame emission spectrometer.

fluids), soils, plant materials, plant nutrients, and samples of inorganic substances. The quantitative estimation of most metals, without concentration, is in the range of 0.1 ppm to 2 ppm.

Detectability limits are given in the following table:

Element	ppm
Barium	0.03
Calcium	0.005
Cesium	0.005
Gallium	0.01
Indium	0.03
Lanthanum	0.1
Lithium	0.00003
Potassium	0.001
Rubidium	0.002
Sodium	0.0001
Strontium	0.004

Limitations

Since flame emission spectrometry deals with the emission of radiant energy, it is concerned with the types of spectroscopic interferences that are found with other emission techniques. Spectral overlap may be caused by the overlap of discrete emission lines or by the band emission of a molecular species. Attempts to reduce this overlap would include increasing the spectral resolution, selecting alternate emission lines for the analyte, or by removal of the interfering substance. Ionization of the analyte, which reduces the number of neutral atoms that can be excited, will decrease the intensity of the emission spectrum. Ionization effects can often be overcome by the addition of an easily ionizable element (e.g., potassium, cesium). Solution properties, including acid and salt concentrations, the presence of organic solvents, and the viscosity, can affect the emission intensities in the sample versus the standard solutions. Matching the matrix of the standard solution with that of the sample can often alleviate some of these solution problems. Releasing agents which preferentially combine with the interferent, are also used to solve problems caused by the type of solution being analyzed.

References

(1) Dean, J. A., Flame Photometry; McGraw-Hill: New York, 1960.
(2) Mann, C. K.; Vickers, T. J.; Gulick, W. M., Instrumental Analysis; Harper and Row: New York, 1974.
(3) Willard, H. H.; Merritt, Jr., L. L.; Dean, J. A., Instrumental Methods of Analysis; 4th edition, Van Nostrand: Princeton, NJ, 1965.

INDUCTIVELY COUPLED ARGON PLASMA EMISSION SPECTROSCOPY

Use

Inductively coupled argon plasma emission spectroscopy (ICP), like atomic absorption methods, is a technique for determining the concentration of elements in solution. The chief advantage of ICP is its ability to analyze many elements, either simultaneously or in a rapid sequential manner depending upon the type of instrument employed.

Sample

The sample may be either liquid or solid, but solid materials must undergo a suitable preparation involving dissolution, decomposition, or extraction. The quantity required for analysis will depend upon the anticipated concentration level and the sensitivity of the element sought. Generally, several grams of material is sufficient for preparation and analysis.

Principle

ICP spectroscopy is based upon the principle that the energy of emission is specific for each element. The liquid sample is atomized by a nebulizer into a stream of argon gas, which carries the atomized sample into the plasma where the elements in solution are thermally excited. The plasma is created by a stream of argon gas, which is ionized when it passes through an RF magnetic field. The excited elements emit photons which are detected by one or more photomultiplier tubes, depending upon the type of instrument. One instrument, a simultaneous system, utilizes photomultiplier tubes (PMT) positioned at predetermined wavelengths on a focal curve with one PMT for each element. A sequential system has a single PMT with a computer-controlled grating that rotates to focus preprogrammed regions of the spectrum on the exit slit. The intensity of the signal is proportional to the concentration of the element. A schematic of a typical sequential ICP instrument is given in Figure 1.

Applications

ICP spectroscopy can be utilized in the same problem areas as atomic absorption spectrosopy. Recent emphasis has been on the analysis of alloys, ceramics, polymers and pickling baths. The ICP detection limits are similar to those of flame atomic absorption methods except for the refractory elements, (e.g., boron, zirconium, and phosphorus) which are more readily determined in the hotter plasma source.

Probably the single most important advantage of ICP is its multielement capability. Using three solutions, the multielement standard, the sample, and a blank, a sequential ICP can rapidly analyze a sample for any number

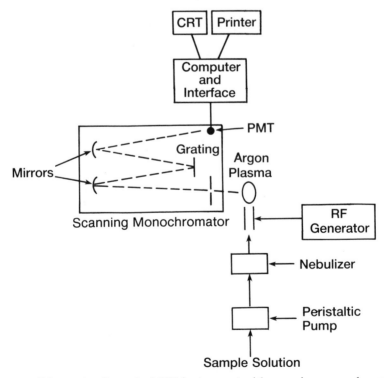

Figure 1. Schematic of a typical ICP instrument with scanning monochromator.

of elements (Figure 2). The multielement capability is further aided by quantitation over a wide concentration range, making the determination of major and trace components possible without the need for dilution.

Another type of ICP instrument, ICP-MS using the argon plasma as a source of singly charged ions and a mass spectrometer as the detector, has been introduced recently. The interface between the argon plasma and the quadrupole mass spectrometer consists of several regions of reduced pressures. Gas from the plasma enters through a metal cone into an area with a pressure of 1 torr. The ions then enter a second chamber (pressure 10^{-4} torr) through a conical skimmer and are scanned by a quadrupole mass analyzer, which is capable of unit resolution over a mass range of at least m/z = 300. Detection limits for ICP-MS are generally several orders of magnitude lower than the corresponding limits for ICP-atomic emission spectrometry. The linear dynamic range has been measured at up to 5 orders of magnitude. ICP-MS has the unique capability to determine isotope ratios for elements in solution, and it can be used to monitor isotope enrichment experiments.

Limitations

Interferences, which may occur during ICP analyses, can be categorized as spectral overlap or matrix effects. Spectral overlap, the direct overlap of the emission lines from the analyte and the interfering element, can be avoided by choosing an alternate line. This is easily accomplished in a sequential instrument, but an alternate line may not be available in a simultaneous system. Matrix effects have also been shown to be a source of interferences. These include changes in acid concentration or dissolved solids

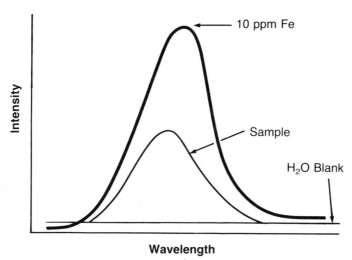

Figure 2. Superimposed emission spectra for semiquantitative analysis of iron.

content, both of which can alter the efficiency of nebulization and hence the sensitivity.

References

(1) Fassel, V. A., Science, 202, 1978, 183.

(2) Fassel, V. A.; Kniseley, R. N., Analytical Chemistry, 46, 1974, 1110A.

(3) Thompson, M.; Walsh, J. N., A Handbook of Inductively Coupled Plasma Spectrometry; Blackie and Son Limited: Glasgow, 1983.

NEUTRON ACTIVATION ANALYSIS

Use

Neutron activation analysis is one of the most sensitive techniques available for trace and ultratrace elemental analysis. Approximately 75 elements can be determined in all types of inorganic and organic samples including coal, oil, polymers, glasses, biological tissue, foods and environmental pollutants. Detection limits of 10^{-9} grams are often attainable.

Sample

Sample preparation is minimal. Liquids and solids in quantities as low as 1 mg can be analyzed as received by this non-destructive technique. Usually 1–10 grams of material are irradiated. Vapor samples must be adsorbed on an inert support such as activated charcoal.

Principle

The sample on exposure to a flux of neutrons forms radioactive species which differ in radiation properties such as half-life, type of radiation emitted, and energy of the radiation. Measurements of one or more of these properties identifies the element of interest.

Quantitative analysis is attained by counting the radiation of the sample and a standard that has been irradiated under identical conditions. For ultratrace analysis, it may be necessary to chemically separate the element of interest from the interfering radioactive matrix. Irradiation time can vary from several minutes to 8–10 hours. Precision is dependent on the radiation counting statistics and is about 1%.

Applications

Neutron activation analysis can be applied to every type of material. Some applications are:

1. Oxygen, nitrogen and fluorine in polymers and other organic compounds.
2. Trace metals in high purity materials such as metals and glasses for optical fiber applications.
3. Mercury, chromium, cadmium and zinc in fish and other foods.
4. Trace metal residues in catalysts and lubricating oils.
5. Scan for 39 elements in unknown samples such as rocks, ashes, soil sediments, and coal.

The neutron activation technique is extremely sophisticated. It requires a nuclear reactor to generate a flux of neutrons (typically, $1.5 \times 10^{13}/cm^2/sec$), high resolution gamma ray counting equipment and specially trained personnel to perform the analyses. The service is available at several industrial laboratories and academic institutions such as MIT, Cambridge, MA; Texas A&M University, College Station, TX and North Carolina State University, Raleigh, NC.

Limitations

The technique is limited to solids and liquids. Sensitivity varies considerably among elements. The element detected must yield products with reasonable half-lives and measurable radiation. The matrix must not form the same product as the element to be detected or isotopes whose radioactivity masks the trace element of interest. Quantitative analysis may require a chemical separation from the matrix after irradiation. The loss of elements during sample dissolution or subsequent chemical separation is a major concern.

References

(1) Brune, D.; Forkman, B.; Persson, B. Nuclear Analytical Chemistry; Verlag Chemie: Weinheim, Fed. Rep. Ger., 1984, 557.
(2) Das, H. A.; Foanhof, A.; Van der Slott, H. A. Studies in Environmental Science, Vol. 22, Environmental Radioanalysis; Elsevier: Amsterdam, Neth., 1983; 298.
(3) Heydon, K., Neutron Activation Analysis for Clinical Trace Element Research; Vol. I and II, CRC Press: Boca Raton, FL, 1984; 172, 217.

CHAPTER 6

X-Ray Analysis

Franz Reidinger, N. Sanjeeva Murthy, Steven T. Correale

X-RAY POWDER DIFFRACTION

Use

X-ray powder diffraction is used to obtain information about the structure, composition and state of polycrystalline materials. The samples may be powders, solids, films or ribbons.

Samples

The minimum amount of material required is a few milligrams. However, greater accuracy is achieved if up to a gram of the sample is available. In order for a solid sample to be mounted in an automatic sample changer, there are restrictions on its dimensions, depending on the instrument used.

Principle

If a beam of monochromatic x-radiation is directed at a crystalline material one observes reflection or diffraction of the x-rays at various angles with respect to the primary beam, (see Figure 1). The relationship between the wavelength of the x-ray beam, λ, the angle of diffraction, 2θ, and the distance between each set of atomic planes of the crystal lattice, d, is given by the well known Bragg condition

$$n\lambda = 2d \sin \theta$$

where n represents the order of diffraction. From this equation we can calculate the interplanar distances of the crystalline material being studied. The

A: Collimation Assembly
B: Sample
C: Slit
D: Exit Beam Monochromator
E: Detector
X: Source of X-Rays

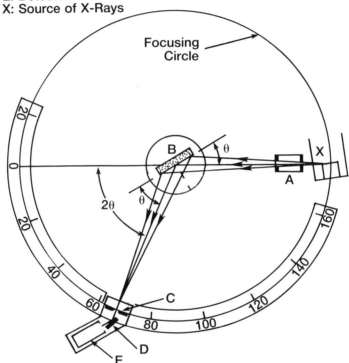

Figure 1. Schematic of an x-ray diffractometer.

interplanar spacings depend solely on the dimension of the crystal's unit cell while the intensities of the diffracted rays are a function of the placement of the atoms in the unit cell.

Applications

The x-ray pattern of the crystalline substance can be thought of as a "fingerprint", each crystalline material having, within limits, a unique diffraction pattern.

The Joint Committee on Powder Diffraction Standards has published the powder diffraction patterns of some 50,000 compounds. An unknown compound is identified by comparing the interplanar spacings and intensities of its powder pattern to the patterns in the powder diffraction file. If x-ray fluorescence data, which describe elemental composition are added, the number of patterns can be reduced. A systematic search (by computer) usually leads to an identification in about one hour. Mixtures of up to nine compounds can often be completely identified. The minimum limit of detection for a single phase in a complex mixture is about 1%. Figure 2 shows the final step of a typical phase identification in which the stick diagrams of the stored compounds are being compared with the diffraction pattern of the sample. In this example αAl_2O_3 (corundum) could be identified as the major crystalline impurity in the preparation of AlN.

In addition to identifying the compounds in a powder, analysis of the diffraction pattern is also used to determine crystallite size, the degree of crystallinity of rapidly solidified materials, the phase composition of the surface region of transformation-toughened ceramics, and other parameters associated with the state of crystalline materials. The diffraction pattern of a solid piece of yttria partially stabilized zirconia is shown in Figure 3. The phase composition has been determined quantitatively by an on-line least squares program which permits the separation of overlapping peaks.

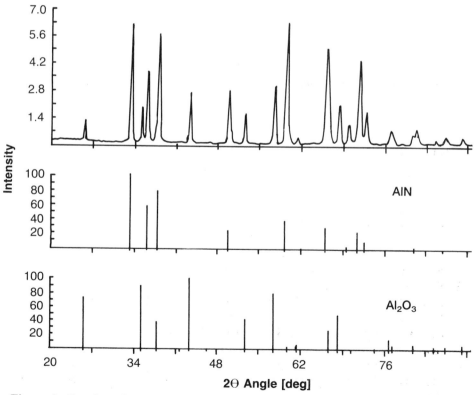

Figure 2. Results of computer-assisted phase identification of a two-component system.

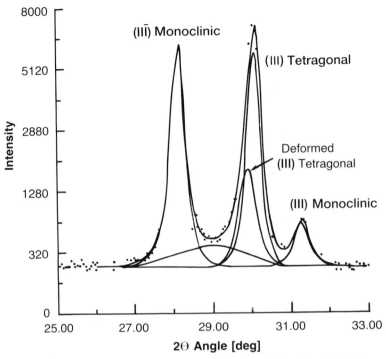

Figure 3. Deconvolution of tetragonal and monoclinic phases in a $Zr(Y)O_2$ ceramic.

Other Specific Uses

A few of the specialized uses of x-ray diffraction are listed below:

1. Low and high temperature diffractometry (between -100 and $1600°C$).
   a. To determine thermal expansion coefficients.
   b. For phase diagram and phase transition studies.
   c. To study order-disorder transitions.
2. Determination of precise crystallographic lattice constants (to ±0.0001 Å), in particular for solid solutions, i.e., the incorporation of atomic or molecular species into a host lattice without change of symmetry.
3. Structure determination of moderately complex structures by the refinement of the complete XRD pattern according to Rietveld. In many cases the results of this technique rival in accuracy a typical single crystal structure determination. It may be required, however, that preferred orientation be eliminated by spray drying techniques.
4. The capability to study the state of microcrystalline and amorphous solids. This is accomplished by computer analysis of time-averaged x-ray data. The results can be used to study the spatial relationship or arrangement of atoms in non-crystalline solids.

Limitations

Film methods are employed to complement the powder diffraction data for the study of materials with preferred orientation. More detailed characterization may require specialty instruments such as a thin film camera, texture goniometer or a double crystal diffractometer. The lack of spatial resolution can be overcome by microdiffraction for polycrystalline samples and x-ray topography for single crystals.

References

(1) Azaroff, L. V., Elements of X-Ray Crystallography; McGraw Hill Book Company: New York, 1968.

(2) Barrett, C. S.; Cohen, J. B.; Faber, J. Jr.; Jenkins, R.; Leyden, D. E.; Russ, J. C.; Predecki, P. K., Advances in X-Ray Analysis; Vol. 29, Plenum Press: New York, 1986.

(3) Klug, H. P.; Alexander, L. E., X-Ray Diffraction Procedures for Polycrystalline and Amorphous Materials; Wiley: New York, 1974.

Acknowledgment

Figure 1 is reprinted from: Cullity, B. D., Elements of X-Ray Diffraction; 2nd Ed., Addison-Wesley Publishing Company, Inc.: Reading, Massachusetts, 1978; 189; with permission of Addison-Wesley Publishing Co.

X-RAY DIFFRACTION—POLYMERS

Use

X-ray diffraction data from polymers provide information about crystallinity, crystallite size, orientation of the crystallites and phase composition in semicrystalline polymers. With appropriate accessories, x-ray diffraction instrumentation can be used to study the phase changes as a function of stress or temperature, to determine lattice strain, to measure the crystalline modulus, and with the aid of molecular modeling to determine the structure of polymers. With a two-dimensional position sensitive detector, temperature and stress dependent measurements, and studies of phase transitions as a function of heat-flow can be carried out in real-time.

Sample

The samples can be films, fibers, plaques, pellets, powders, and even gels. Usually 100 mg to 1 gm of material is needed. The thickness of the film or plaque can vary from 25 μm to 1 mm, and be at least 2 cm × 2 cm in area. Under some circumstances films of size 1 mm × 2 mm can also be analyzed. Although fibers as short as 5 mm can be analyzed, usually a length of 1 m is necessary.

Principle

A diffraction pattern is a distribution of scattered intensity as a function of scattering angle. The x-ray diffraction (XRD) pattern of a hypothetical completely crystalline polymer would be a series of sharp peaks, each corresponding to diffraction (reflection) from one of the various crystallographic planes. The XRD pattern of a completely amorphous polymer is a broad halo which represents the average separation of polymer chains. The XRD pattern of a semicrystalline polymer is then a superposition of crystalline reflections on an amorphous halo. An index of crystallinity can be obtained from the ratio of the integrated intensity of the crystalline peaks to the total area under the XRD curve. The crystalline peaks contain information about the unit cell dimensions, presence of various polymorphs, crystallite size, and lattice strain. The degree of crystalline orientation is obtained from the azimuthal spread of the reflections. From the changes in unit cell dimensions as a function of stress, and by using suitable models, one can calculate the crystalline modulus. Since only a small number of reflections are obtained from a typical polymer, the structure is usually determined by trial and error using molecular modeling methods.

Applications

Figure 1a shows an equatorial scan of a typical nylon 6 fiber resolved into α crystalline peaks, γ crystalline peaks and an amorphous halo. The inset shows the x-ray diffraction photograph of the fiber. From such data one can determine the crystallinity, relative fraction of the α and γ forms, and crystalline orientation. Figure 1b shows a detailed profile analysis on high-angle equatorial x-ray diffraction data from the α form of nylon 6. Results from such an analysis are essential for an accurate determination of unit cell parameters. Profile analysis is accomplished by performing a non-linear least squares analysis using a Pearson VII function.

Figure 1. Equatorial scan of a nylon 6 fiber. Figure a shows the pattern from 15° to 30° 2θ resolved into α, γ, and amorphous components. The α form has two reflections at ~20.3° and at ~23.5°; the γ form has two reflections at ~21.3° and at ~ 22.5°. The inset shows a typical x-ray diffraction photograph of a fiber. Figure b shows the higher angle equatorial scan of an α-rich fiber resolved into six individual reflections.

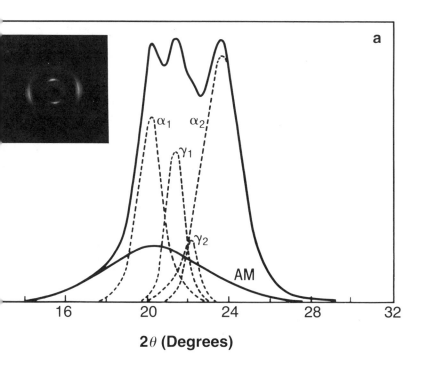

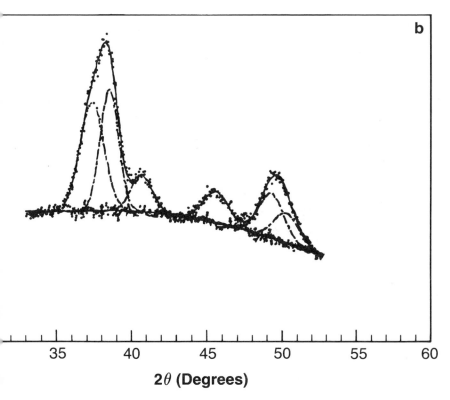

Figure 2 shows the structure derived from an alkali-metal complex of polyacetylene using a trial and error method. The chain-axis projection shows columns of alkali-metal ions in the channels defined by four polyacetylene chains.

A two-dimensional position sensitive detector is used to study time dependent phenomena such as desorption in iodine-polymer complexes and phase changes in polymers due to stress and heat-flow. Feasibility of such analysis is illustrated in Figure 3 which is an x-ray diffraction pattern of an oriented HALAR® filament. Equatorial reflections, representing packing of the chains, appear as two bright spots. The diffuse meridional streaks along the fiber axis can be used to evaluate the ordering of the polymer chains in the chain-axis direction.

Limitations

The x-ray diffraction technique is useful for characterizing the ordered state of matter, but interpretation of the data from disordered/amorphous fractions in a specimen is not always straight forward. XRD measurements in polymers usually represent contributions from structures over ~1 mm sampling depth. Therefore, surface structures cannot be easily analyzed, and structures in multilayer samples cannot be readily separated. Although XRD can be used to confirm the existence of a given polymer, and even a

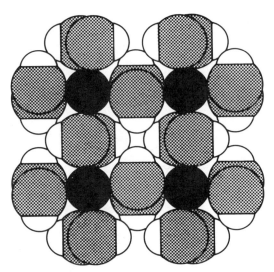

Figure 2. Chain-axis projection of the structural model for alkali-metal doped polyacetylene. Large dotted circles are carbons, and small open circles are hydrogens. The metal ions are shown as filled circles.

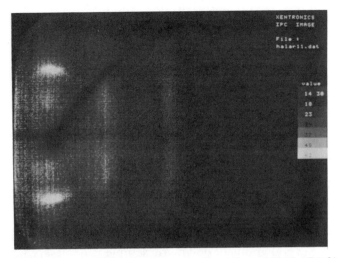

Figure 3. Wide-angle x-ray diffraction pattern of an oriented HALAR® filament. The fiber axis is horizontal. The detector was positioned at 30° 2θ. The pattern was obtained in two minutes at 30kV and 20 mA.

particular phase of the polymer, it cannot always identify a previously unknown polymer. Analysis of XRD data from blends is still in developmental stages.

References

(1) Alexander, L. E., X-Ray Diffraction Methods in Polymer Science; John Wiley and Sons: New York, 1969.

(2) French, A. D.; Gardner, K. H., eds.; Fiber Diffraction Methods; American Chemical Society: Washington, DC, 1980.

(3) Tadokoro, H., Structure of Crystalline Polymers; John Wiley and Sons: New York, 1979.

Acknowledgments

(1) Figure 1 is reprinted from: Murthy, N. S., American Laboratory, 14, No. 11, 1982, 72; with permission of International Scientific Communications, Inc.

(2) Figure 2 is reprinted from: Baughman, R. H.; Shacklette, L. W.; Murthy, N. S.; Miller, G. G.; Elsenbaumer, R. L., Mol. Crys. Liq. Cryst., 118, 1985, 256; with permission of Gordon and Breach, Inc.

SMALL-ANGLE X-RAY AND NEUTRON SCATTERING

Use

Small-angle scattering with x-rays and neutrons (SAXS and SANS) is used to study structures in the range of 10–1000 Å. Small-angle neutron scattering (SANS) is especially useful in studying the distribution of a particular polymer in a blend.

Sample

For small-angle x-ray scattering (SAXS) studies of solutions, ~0.2 ml of solution at ~5% concentration is needed. 100 mg of solid sample, 0.5 mm $\times$ 1 cm $\times$ 1 cm polymer films or plaques, and fibers of ~20 cm length are usually used. Several grams of polymeric sample is necessary for SANS.

Principle

X-ray scattering at low-angles reflects the electron-density fluctuations and contrast which might exist over a range of 10 to 1000 Å. In a semicrystalline polymer, the electron density difference between the crystalline regions and the interlamellar amorphous regions provides information about the lamellar structure. With polymers in solution, or porous materials, the SAXS curve is determined by the size and shape of the polymer, or the size distribution of the voids. SANS data obtained by labeling one polymer in a blend with deuterium, can be used to study the distribution of this labeled molecule in the matrix.

Applications

Typical applications of both SAXS and SANS techniques include determination of the size of globular particles (proteins, polymers, inorganics) in solution or suspension, and void size analysis in ceramics and catalysts. A schematic of a typical SAXS apparatus is shown in Figure 1. The diffuse SAXS from three samples of globular BOC-polylysine molecules dissolved in N,N-dimethyl formamide is shown in Figure 2. Note the sharp fall off of the scattered intensity for the large molecules and the weak scattered intensity for the smaller molecules. SAXS and SANS are also used to study lamellar structure in semicrystalline polymers. Figure 3 shows the 80 Å lamellar repeat from a nylon 6 film. A correlation has been established between this

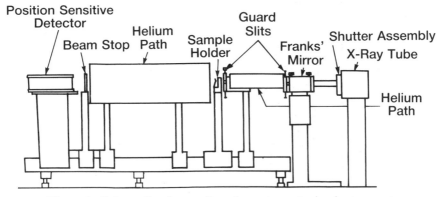

Figure 1. Schematic of a small-angle x-ray scattering instrument.

lamellar repeat and heat-setting temperatures in nylon 6 carpet-fibers. Absence of such a lamellar repeat suggests the presence of extended chain molecules.

SANS has also been used to study the distribution of water molecules (D_2O) in nylon 6 films of varying crystallinity. In another study, by deuter-

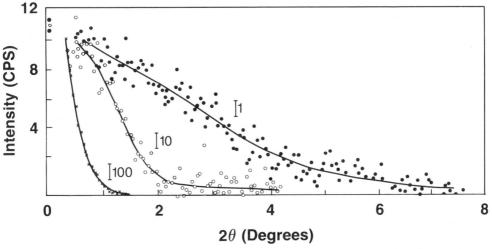

Figure 2. Diffuse scattering from three samples of globular BOC polylysine molecules dissolved in N,N-dimethyl formamide; scattering from samples, after subtracting the solvent background, of three molecular weights, 1,900 (●), 29,000 (○), and 233,600 (x) obtained respectively in 16 hours, 1,800 seconds and 1,200 seconds are shown. Concentrations are 5% and the intensity scales are given next to the vertical bars near each of the curves. Radii of gyration calculated from such curves after extrapolation to zero concentration are 8.0, 19.9 and 43.4 Å, respectively.

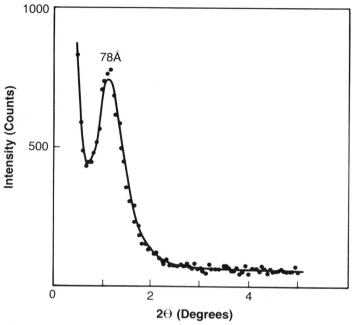

Figure 3. Small-angle x-ray scattering curve from a semicrystalline nylon 6 film. The peak at 78 Å represents the distance between the crystalline lamellae.

ating polyethylene terephthalate (PET), it has been shown that PET and polyester carbonate are miscible in the amorphous phase.

Limitations

The technique is versatile for comparative evaluation of materials, but an absolute evaluation of the data is difficult. The interpretation of the results is dependent on the model being adopted. The measurements are time-consuming and may last up to 24 h, thus the stability of the sample during such measurement is of some concern. Analysis of polydisperse samples is not always unambiguous. Phase segregation of deuterated fractions may make SANS less attractive for some polymers.

References

(1) Glatter, O.; Kratky, O., Small-Angle X-Ray Scattering; Academic Press: New York, 1982.
(2) Guinier, A., X-Ray Diffraction: Freeman & Co.: San Francisco, 1963, pp 319–350.
(3) Higgins, J. S., J. Appl. Cryst., 11, 1978, pp 346–375.

Acknowledgments

Figures 1, 2 and 3 are reprinted from: Murthy, N. S., Norelco Reporter, 30, 1983, 39; with permission of Philips Electronic Instruments, Inc.

X-RAY ABSORPTION SPECTROSCOPY

Use

In addition to its limited use as a tool for elemental analysis, comparatively recent theoretical advances have shown that x-ray absorption spectroscopy yields two types of information:

1. EXAFS (Extended X-Ray Absorption Fine Structure Spectroscopy) determines the radial distribution function, i.e., the type, number and distance of neighboring atoms of the particular atomic species whose absorption spectrum is being obtained. This provides considerable information about the molecular structure, which is particularly useful in studying intermetallic and organometallic compounds.
2. XANES (X-Ray Absorption Near Edge Spectroscopy) yields information regarding the nature of unoccupied electronic states of the atom in question. This in turn can be related to electrical behavior, magnetic behavior, density of states, stereochemistry, etc. The precise position (energy) of the absorption edge can give the ionicity or charge (i.e., oxidation state) of the atom in question.

Sample

For the two types of experiments the sample requirements are the same. Samples are usually mounted as thin foils, as compressed pellets or simply dusted onto tape. Liquids are enclosed in suitable containers of plastic or quartz. The only requirement for sample mass is that the x-ray beam be uniformly attenuated by a certain thickness which depends on the elemental composition of the sample, especially on the concentration of the element in question. The optimal thickness, T, is obtained if the product $(\mu_H - \mu_L)T$ is about one, where μ_H and μ_L are the absorption coefficients above and below the absorption edge, respectively.

Principle

Although the experimental equipment required is complex, the measurement is straightforward. The specimen is placed in a beam of x-rays and the incident and transmitted intensities are monitored as the energy of the x-rays is varied. From these data a plot of the absorption edge (see Figures 1 & 2) and of the high energy side of the edge is obtained (see Figures 3 & 4).

The position (energy) of the absorption edge is somewhat analogous to "chemical shifts" in other types of spectroscopy and is affected by the net charge on the atom. Thus, oxidation state information is obtained.

The structure below and at the edge (see Figures 1 & 2) represents transitions of core electrons to higher energy states of the parent atom. Hence, information about the normally unoccupied energy states of the atom in a particular matrix is obtained.

The fine structure (EXAFS) on the high-energy side of the absorption edge (see Figures 3 & 4), occurs because the electron completely ejected from the atom in question is backscattered by its neighboring atoms. The Fourier transform of these spectral features yields a qualitative representation of the radial distribution function while its precise parameters are determined by least squares analysis, i.e., the distance, number and type of neighboring atoms and an estimate of the disorder affecting the interatomic distance.

Applications

X-ray absorption spectroscopy can be applied to characterize:

1) Catalysts
2) Amorphous materials

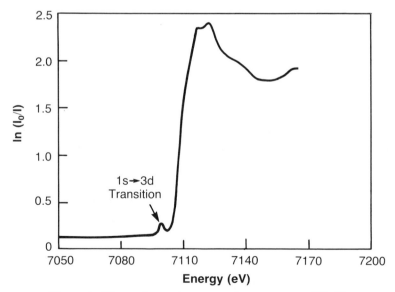

Figure 1. X-ray absorption near edge structure of $FeCl_3$.

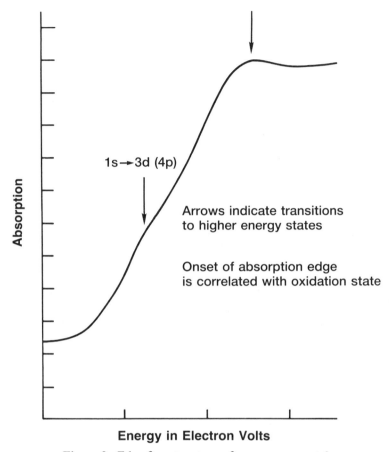

Figure 2. Edge fine structure of manganese metal.

3) Intercalation compounds
4) Semiconductors
5) Clusters
6) Surface-substrate interactions
7) Electrochemistry
8) New molecules
9) Ionomers
10) Behavior of molecules in solution

Limitations

The main shortcoming of XANES is the lack of a readily applicable quantitative theory to correlate data and structural model. EXAFS suffers from the lack of low momentum data. Therefore, in moderately disordered sys-

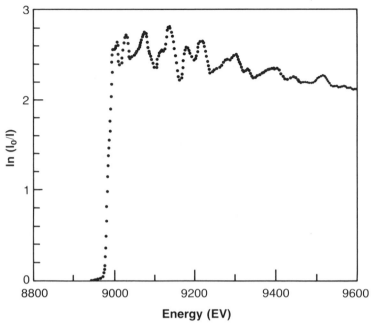

Figure 3. EXAFS spectrum of Cu.

tems information on second and higher order shells may be absent, and in severely disordered systems even the parameters of the first order shell may be affected.

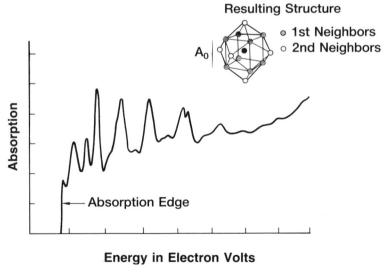

Figure 4. EXAFS spectrum of iron.

References

(1) Hodgson, K. O.; Hedman, B.; Penner-Hahn, J. E., eds.; EXAFS and Near Edge Structure III; Springer-Verlag: Berlin, 1984;.

(2) Teo, B. K.; Joy, D. C., eds.; EXAFS Spectroscopy: Techniques and Applications; Plenum Press: 1981.

X-RAY FLUORESCENCE SPECTROSCOPY

Use

X-ray fluorescence is a relatively simple and, in general, nondestructive method for the analytical determination (both qualitative and quantitative) of elements. This method is extremely useful because of the ease of sample preparation and its ability to "scan" the periodic table for all elements down to aluminum and under certain circumstances to boron.

Sample

For qualitative determination of the elements present in a sample there is virtually no sample preparation. That is, the sample can be run "as received". It can be a liquid, solid, emulsion, etc. The sample can be several centimeters in diameter or weigh as little as a few mg.

For quantitative analysis of any or all of the elements present in a sample, virtually no preparation is required for liquids. For solids some preparation is required such as mixing with a fluxing agent to insure sample homogeneity.

Principle

The x-ray fluorescence (XRF) technique is based on the principle that if an atom is bombarded with highly energetic photons, some of its electrons are ejected. As other electrons fill the energy levels vacated by the ejected electrons, they emit quanta of radiation characteristic of that particular atom type. Hence, each element has its own set of characteristic emission or x-ray fluorescence lines.

There are two types of experimental equipment available. One is based on wavelength dispersive x-ray fluorescence (WDXF), the other is based on energy dispersive x-ray fluorescence (EDXF). They differ only in the manner the emitted radiation from the sample is dispersed. WDXF uses a crystal grating to separate the energies and a conventional x-ray detector to measure the radiation (see Figure 1). EDXF uses a solid state detector which has an inherent ability to separate the energies without any mechanical movement. EDXF has the advantage of speed but the disadvantage of poorer sensitivity and resolution.

The most recent developments in XRF have been the incorporation of minicomputers with the spectrometers and the development of fundamental parameters software. The minicomputers permit automated data collection and data reduction using quantitative and matrix correction algorithms. The fundamental parameters software attempts to predict theoretically the XRF spectrum for a given composition using a minimum number of standards (which may be pure metals and/or oxides). The program will refine its spectrum and predict the composition of the sample. As with all XRF quantitative analysis, the accuracy of results depends on how closely the standard(s) resemble the sample.

Applications

a) Rapid qualitative determination of the elements present in an unknown material with virtually no sample preparation and semi-quantitative determination by using simple correction factors.
b) Detection of all elements down to B in the periodic table from a lower limit of detection (LLD) of a few ppm (see Figure 2) up to 100%.
c) Quantitative determination of any or all the elements in a sample (excluding those elements less than 5 in atomic number).
d) Surface (~ 1 μm) composition comparison with the bulk (>10 μm) using emission lines with different depth of penetration for certain samples (ceramics).

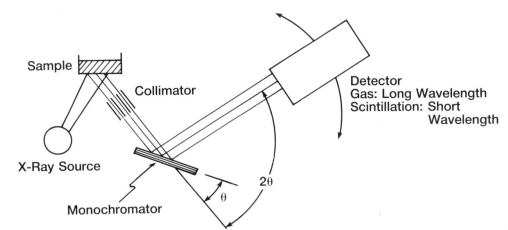

Figure 1. Schematic of a wavelength dispersive x-ray fluorescence spectrometer.

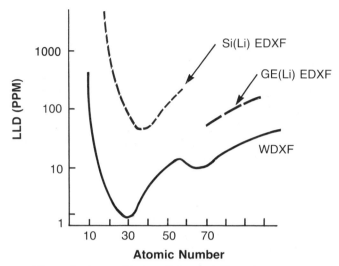

Figure 2. Lower limits of detectability of elements.

Limitations

The primary limitation of XRF is the lack of sensitivity to elements lighter than Al. Another disadvantage is that for accurate quantitative analysis usually standards are required which are similar in composition and morphology to the unknown. However, the advent of fundamental parameters software has begun to alleviate this limitation.

References

(1) Bertin, E. P., Principles and Practice of X-Ray Spectrometric Analysis; Plenum Press: New York, 1975.

(2) Jenkins, R.; Gould, R. W.; Gedcke, D., Quantitative X-Ray Spectrometry; Marcel Dekker, Inc.: New York, 1981.

(3) Tertian, R.; Claisse, F., Principles of Quantitative X-Ray Fluorescence Analysis; Heyden and Sons Inc.: Philadelphia, 1982.

SYNCHROTRON X-RAY SOURCES

Use

Synchrotron radiation sources provide an intense x-ray beam which is not available with in-house x-ray instrumentation. The intense x-ray beam allows many specialized x-ray analyses to be performed.

Principle

The brilliance of in-house x-ray sources is limited by the poor efficiency of the x-ray generating process and the limited rate with which heat can be carried away from the target. This restriction does not apply to x-rays generated as synchrotron radiation by high energy electrons in an accelerator. The spectrum of such a source extends continuously over a wide range of energies (see Figure 1). The high flux can best be exploited by an arrangement which lets the "white" beam interact with the sample and uses a solid state detector as an analyzer. This set-up is especially suited for microdiffraction. In the second, more common arrangement, a very narrow energy band of the spectrum is selected by a double monochromator which consists of perfect Si or Ge crystals. This set-up provides x-rays of continuously tunable energy corresponding to wavelengths of 0.7 to 3 Å with an energy distribution as narrow as the CuK_α peak and an angular divergence of about 0.001 degrees. The intensity of this beam is about ten times that of the CuK_α radiation of a standard tube. In the near future intensity increases of up to a thousand are expected following the installation of wigglers and undulators.

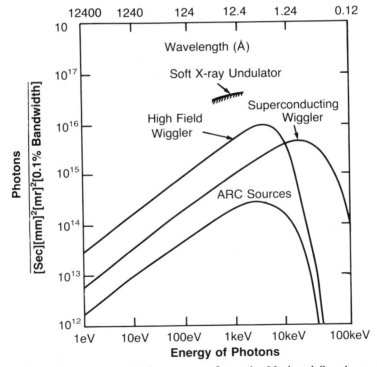

Figure 1. Spectra of x-ray radiation sources from the National Synchrotron Light Source at Brookhaven National Laboratory.

Applications

a) EXAFS and XANES (see section on X-Ray Absorption Spectroscopy)
b) Rocking curves
c) Preferred orientation
d) Residual stress
e) Depth profiling of phase composition
f) Microdiffraction
g) X-ray topography

Limitations

Time constraints, scheduling problems for different techniques and fatigue caused by a 24 hour working day are the main limitations that affect an industrial user. Time pressure and fatigue are detrimental to a high success rate especially if an in-house instrument for preliminary screening is not available and critical decisions have to be made at odd hours.

References

(1) Bienenstock A.I.; Winick, H., Physics Today, 36, 1983, pp 48–58.
(2) Winick, H.; Doniach, S. eds.; Synchrotron Radiation Research; Plenum Press: New York, 1980, 775.
(3) White-DePace, S.; Gmur, N. eds.; National Synchrotron Light Source Annual Report; BNL-Associated Universities Inc.: Upton, New York, 1986.

Acknowledgment

Figure 1 is reprinted from: NSLS Annual Report 1984; with permission of Brookhaven National Laboratory.

CHAPTER 7

Microscopy

Annemarie C. Reimschuessel, John E. Macur, Jordi Marti

OPTICAL MICROSCOPY

Use

Optical microscopy is used for the examination and characterization of matter using visible light.

Sample

Any solid material and many liquids may be studied.

Principle

In optical microscopy, information is obtained by light transmission through or reflection from matter. Optical microscopes consist of a light source, a condenser, two lens systems and other accessories. They are capable of producing magnified images from 10X through 1400X, thereby permitting observation of structures too small for unaided visual observation. The limit of resolution is approximately 0.2 μm. Camera attachments permit photographic recording of the image. Studies can be conducted using transmitted and reflected light, polarized light, bright field, dark field, differential interference contrast, and phase contrast illumination.

A schematic of a compound optical microscope is given in Figure 1. Thermal microscopy can be conducted at temperatures ranging from $-20°C$ to $350°C$ with the aid of a thermal stage. This allows phase transitions of various materials to be directly observed and transition temperatures measured.

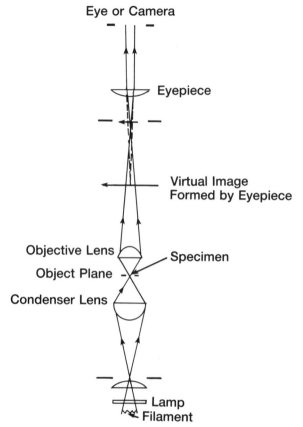

Figure 1. Schematic of a compound optical microscope.

Applications

Transmission optical microscopy is routinely used for studying many types of materials. Some examples are: polymer films or fibers, biological or petrographic thin sections, foam cellular structures, liquids, dispersions, powders and emulsions. By its very nature, the transmission method requires thin (several microns) specimen sections which can be prepared in a number of ways (see Microscopy Specimen Preparation Techniques section). Polarizing elements along the optical path permit the observation of birefringent differences, thickness differences or orientation variations within the sample. For example, Figure 2 shows a cross-section of a defect-containing laminate consisting of five layers (HDPE/nylon/HDPE/nylon/ HDPE). The defect itself is a gel in the high density polyethylene (HDPE) layer. Note the disruption of the layers in this region. The nylon layers show

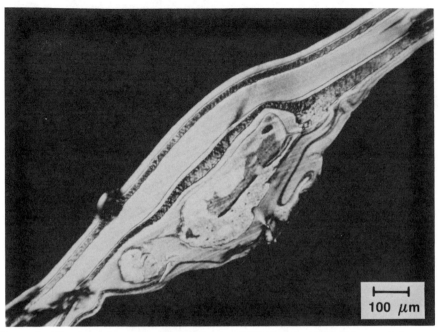

Figure 2. Five layer laminate (HDPE/nylon/HDPE/nylon/HDPE) containing a gel defect shown in crossed polarized light.

a strong spherulitic development. Such optical examination for laminated films is very important, not only for general characterization (number of layers and thicknesses) but also for failure analysis (delamination) in film packages. These samples can also be heated during observation, thereby allowing the polymer transition temperatures to be measured.

Reflected light microscopy is often used in the examination of polished/etched metallurgical or ceramic specimens for inclusion size or grain size determination. Any solid material can be examined using reflected techniques but due to the limited depth of field, the specimen should be flat.

Dark field methods permit the enhancement of surface discontinuities (fissures, raised phases in a smooth matrix, etc.). Phase contrast illumination permits the improved visibility of low contrast objects. This technique is especially useful in the examination of polymer blends; for example, Figure 3 shows phase separation in a nylon copolymer using the phase contrast method.

Optical microscopy is also useful in the study of liquid crystalline materials. Figures 4 and 5 illustrate the morphology of lyotropic liquid crystals. Figure 4 is a photomicrograph of poly (n-octyl isocyanide) in chloroform and Figure 5 is a photomicrograph of poly (ω-decenyl isocyanate) in tetrachloroethane (TCE).

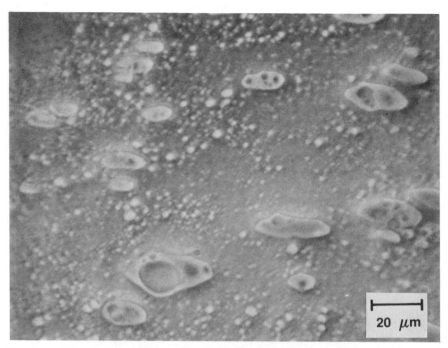

Figure 3. Phase separation in a nylon copolymer using phase contrast microscopy.

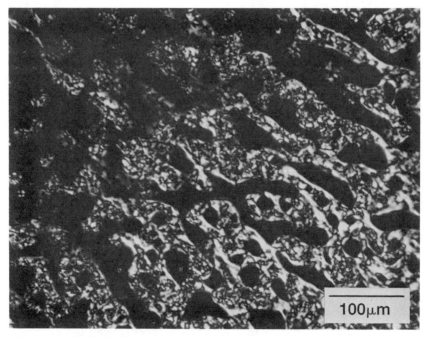

Figure 4. Optical micrograph of poly (n-octyl isocyanide) in chloroform.

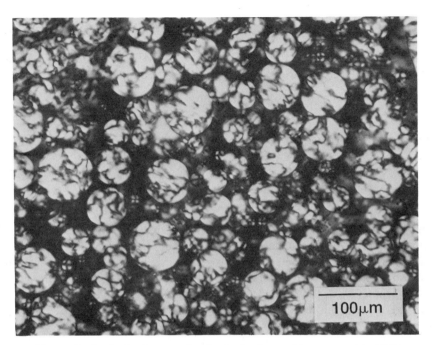

Figure 5. Optical micrograph of poly (ω-decenyl isocyanate) in TCE.

Limitations

When the optical microscope is used in the transmission mode, the specimen must be approximately 5 μm thick; this requires careful specimen preparation, particularly when portions of the specimen are soft and easily deformable. The resolution of the technique is limited to about 0.2 μm. The depth of field varies with numerical aperture and can range from about 8 μm (N.A.-0.25) to about 0.4 μm (N.A.-1.30).

References

(1) Mason, C. W., Handbook of Chemical Microscopy; John Wiley and Sons, Inc.: New York, 1983.

(2) Spencer, M., Fundamentals of Light Microscopy; Cambridge University Press: Cambridge, 1982.

(3) Zieler, H. W., The Optical Performance of the Light Microscope I and II; Microscope Publications Ltd.: London, England and Chicago, Illinois, 1974.

Acknowledgement

Figure 1 is reprinted from: reference 2, p. 5; copyrighted by and with permission of Cambridge University Press.

SCANNING ELECTRON MICROSCOPY (SEM)/ ELECTRON PROBE MICROANALYSIS (EPMA)

Use (SEM)

Scanning electron microscopy is used primarily for the study of surface topography of solid materials. It permits a depth of field far greater than optical or transmission electron microscopy. The resolution of the scanning electron microscope is about 3 nm (30 Å), approximately two orders of magnitude greater than the optical microscope and one order of magnitude less than the transmission electron microscope. Thus, the SEM bridges the gap between the other two techniques.

Sample

Any solid material can be studied. Sample size is limited to specimens less than about 10 cm in diameter.

Principle

An electron beam passing through an evacuated column is focussed by electromagnetic lenses onto the specimen surface. The beam is then rastered over the specimen in synchronism with the beam of a cathode ray tube display screen. Inelastically scattered secondary electrons are emitted from the sample surface and collected by a scintillator, the signal from which is used to modulate the brightness of the cathode ray tube. In this way the secondary electron emission from the sample is used to form an image on the CRT display screen. Differences in secondary emission result from changes in surface topography. If (elastically) backscattered electrons are collected to form the image, contrast results from compositional differences. Cameras are provided to record the images on the display screen.

A schematic of a scanning electron microscope-microprobe analyzer is given in Figure 1.

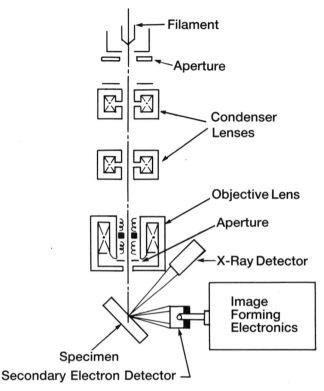

Figure 1. Schematic drawing of a combined scanning electron microscope-microprobe analyzer.

Applications

Scanning electron microscopy has been applied to the study of fibrous materials, ceramics, composites, metals, catalysts, polymers and biological materials. Information may be obtained by examination of both the natural surface of materials and that exposed by either fracture or sectioning. Rough topographic features, void content and particle agglomerations are easily revealed as well as compositional (phase) differences within a material.

Some examples of recent investigations are:

1. Evaluation of textile finishes on fabrics and fibers, and the study of surface change resulting from various fiber treatments. As an example, Figure 2 shows the non-uniform (patchy) distribution of an experimental anti-staining agent on nylon 6 carpet fiber. This non conductive sample demonstrates the novel low voltage (1 keV) imaging capability of the SEM.

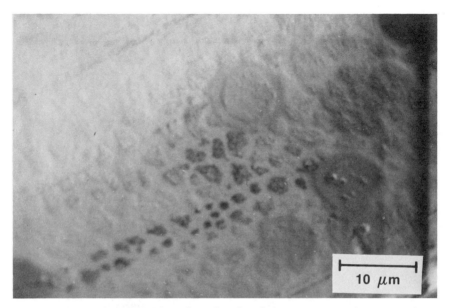

Figure 2. Low voltage SEM imaging of anti-staining agent on nylon carpet fibers.

2. Characterization of optical fiber couplers and connectors; failure mode determination of optical fibers. Figure 3 shows a secondary electron image of the fracture face of an optical fiber. The crystalline inclusions (see insert) were responsible for the failure.

3. Metallographic investigations of amorphous metals and rapidly solidified brazing alloys to study magnetic domains, phase separation and corrosion.

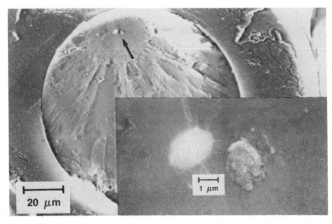

Figure 3. Optical fiber fracture face showing inclusions responsible for failure.

4. Correlations between processing conditions of ceramic materials and their ultimate microstructure.

Use (EPMA)

Electron probe microanalysis is used for determining elemental composition and distribution within a micro volume of material. Elemental detectability extends from B through U. Microprobe analysis is most commonly used in conjunction with SEM imaging thereby allowing analyses to be performed directly on the areas under electron beam observation.

Sample

Any solid material containing elements heavier than Be may be studied.

Principle

When materials are bombarded by a high energy (10 keV-50 keV) electron beam (as in an electron microscope), characteristic x-ray fluorescence radiation is produced. By incorporating either energy dispersive or wavelength dispersive spectrometers directly into the instrument, it is possible to obtain x-ray spectra directly on the area as seen by the electron beam. Thus, it is possible to obtain qualitative and quantitative elemental data from a volume on the order of 1 μm^3 for the elements B through U. Data can be obtained from an isolated region of the sample (spot mode), along a preselected linear trace (line profiling) or from an area (x-ray distribution mapping).

Applications

Because of its capability for qualitative and quantitative analysis on micro volumes, EPMA has been used extensively for impurity and inclusion identification in polymers and cast ingots, as well as compositional determination of amorphous metals, complex garnet films and corrosion deposits. The line profiling and x-ray mapping modes are routinely used for determination of diffusion profiles and elemental distribution in braze joints, electronic components, optical fiber components and plated parts.

Applications of the electron microprobe technique are shown in Figures 4 to 7. The first example (Figure 4) shows a comparison of x-ray spectra that were collected from the inclusion and "control" area of the optical fiber fracture face shown in Figure 3. Note the presence of Y, Sm, Gd, and Dy. This

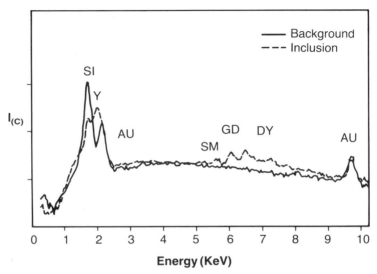

Figure 4. Energy dispersive x-ray spectra indicating that the inclusions shown in Figure 3 consist of rare earth elements, probably as complex oxides.

indicates that the failure resulted from the presence of a complex rare earth oxide inclusion.

Figure 5 shows a backscattered electron image of a multiphase nozzle blockage in cross-section. Lower atomic number phases appear darker. The spherical particles are boron nitride. The right hand portion of the figure shows the B Kα x-ray distribution map of the blockage region shown in the left of the figure. Note the high B concentration in the spherical particles.

Figure 6a shows another example of analysis of multiphase nozzle blockage. Figure 6b reveals the Ca distribution in the blockage area. Figure 7a

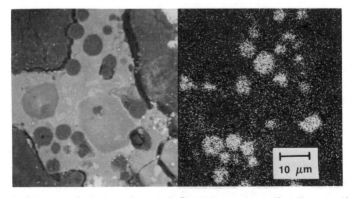

Figure 5. Backscattered electron image (left) and corresponding B x-ray distribution map (right) of a casting nozzle blockage. The dark spheres containing B are hexagonal boron nitride.

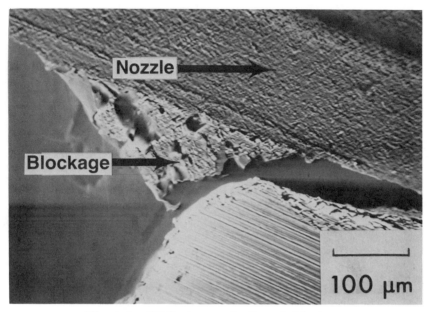

Figure 6a. SEM micrograph of nozzle blockage.

Figure 6b. Ca distribution in the nozzle blockage.

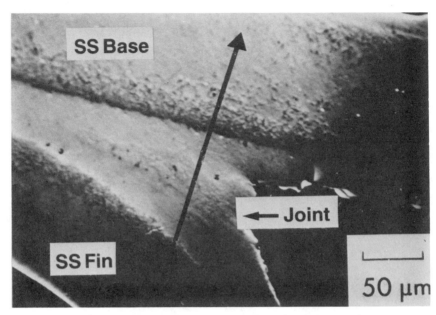

Figure 7a. Braze joint between strips of stainless steel.

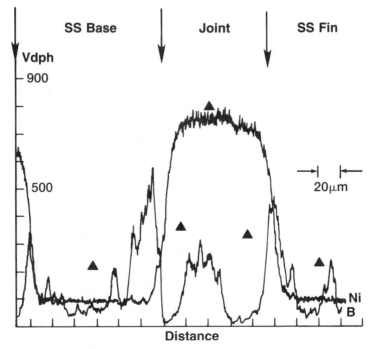

Figure 7b. B and Ni profiles through the braze joint.

illustrates a braze joint between two stainless steel strips. Figure 7b shows the boron and nickel profiles across this joint obtained with the aid of a wavelength spectrometer. Also shown are Vickers hardness values (▲) which correlate with the chemical composition across the joint.

Limitations (SEM)

Samples to be studied must be solids that are not electron-beam-reactive and contain no highly volatile or corrosive components. Special specimen stages can handle specimens up to about 20 cm in diameter although conventional stages can accommodate samples only a few cm in diameter. Resolution limitations of the technique extend down to about 20 Å–30 Å.

Limitations (EPMA)

Due to the fact that microprobe analysis is performed inside an SEM, the above limitations still apply as to sample type and size. Additionally, the specimen surface should be flat and polished. The lightest element detectable is B. Since the electron beam penetration and x-ray emission are strongly affected by instrument parameters, specimen composition and specimen characteristics, the analyzed volume can be as large as several cubic microns. Elemental sensitivity is on the order of 100 ppm for wavelength dispersive x-ray analysis and about 1000 ppm for energy dispersive x-ray analysis. Sensitivity may be poorer depending on the element of interest and it's matrix.

References

(1) Goldstein, J.; Yakowitz, H.; Newbury, D.; Lifshin, E.; Colby, J.; Coleman, J., Practical Scanning Electron Microscopy; Plenum Press: New York, 1975.

(2) Heinrich, K. F., Electron Beam X-Ray Microanalysis; Van Nostrand Reinhold Company: New York, 1981.

(3) Newbury, D.; Joy, D.; Echlin, P.; Fiori, C.; Goldstein, J., Advanced Scanning Electron Microscopy and X-Ray Microanalysis; Plenum Press: New York, 1986.

(4) Wells, O. C., Scanning Electron Microscopy; McGraw-Hill Book Company: New York, 1974.

Acknowledgment

Figure 1 is reprinted from: Goldstein, J. I.; Yakowitz, H., Practical Scanning Electron Microscopy; Plenum Press: New York, 1975, 16; with permission of Plenum Publishing Corporation.

SCANNING/CONVENTIONAL TRANSMISSION ELECTRON MICROSCOPY

Use

Transmission electron microscopy is used to study the structure and morphology of materials by examining the diffracted and transmitted electron intensities. With the scanning attachment the microscope is operated in the STEM (Scanning Transmission Electron Microscope) mode. This is used for image analysis and for elemental microanalysis of areas as small as 50 nm in size.

Sample

Most solid materials can be studied, but because of technical constraints and because of the large scattering of the electrons in the solids, samples are limited to no more than 3 mm in diameter and less than 0.2 μm in thickness. Obtaining a thin sample which will be transparent to the electron beam is a critical aspect of transmission electron microscopy. Various techniques have been developed for this purpose. They include spraying or dusting, sectioning, electropolishing and ion milling.

Particulate materials such as powders or dispersions are often sprayed or dusted onto a thin support film, usually a carbon film mounted on electron microscope grids.

Organic bulk materials such as fibers, films or thicker specimens can be sectioned with an ultramicrotome equipped with a diamond knife. Prior to sectioning, these specimens often need to be embedded in an epoxy resin that cures at 50°C overnight. To minimize distortion, relatively soft specimens are sectioned at low temperatures, typically less than -50°C. The thin sections are mounted on electron microscope grids.

Contrast of organic materials, which is usually low, can be enhanced by staining techniques. The stain selectively defines that portion of the sample in which it resides. Staining substances such as phosphotungstic acid, uranyl acetate and osmium tetroxide are frequently employed.

Thin sections of inorganic bulk substances such as ceramics or metals are obtained by electropolishing or ion milling. A combination of mechanical cutting and thinning techniques followed by ion milling is used to obtain electron transparent areas for the study of cross-sections of semiconductor interfaces.

Principle

In transmission electron microscopy, a beam of high energy electrons (100–400 keV) is collimated by magnetic lenses and passes through a specimen under high vacuum. The resulting diffraction pattern can be imaged on a fluorescent screen below the specimen (see Figure 1). From the diffraction pattern one can obtain the lattice spacings of the structure under consideration. Diffraction information from areas less than 0.1 μm in size can be obtained. Alternatively, one can use the transmitted beam or one of the diffracted beams to form a magnified image of the sample on the viewing screen. These are respectively the bright field and dark field imaging modes, which give information about the size and shape of the microstructural constituents of the material with a resolution of 0.2 nm.

If the incident beam is allowed to raster and the transmitted beam is detected by a scintillator rather than by a photographic plate, a STEM image is obtained on a CRT. The signal from the CRT is coupled to a computer which allows the performance of various functions such as digital imaging, image enhancement and particle size analysis.

As the incident electron beam interacts with the specimen, characteristic x-rays are emitted by the sample. These can then be detected and analyzed. In this way, elemental composition from regions 0.05 μm in size can be obtained.

Figure 2 is a schematic of a scanning transmission electron microscope.

Applications

Characterization of powder samples can be obtained by combining TEM and STEM capabilities. Particle size distributions are obtained from the dig-

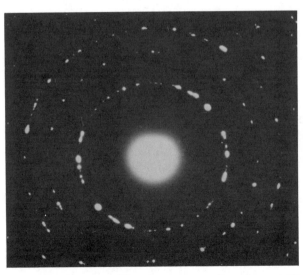

Figure 1. Transmission electron diffraction pattern of a polycrystalline material.

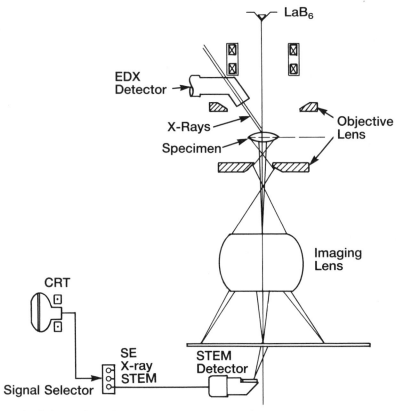

Figure 2. Schematic of (scanning) transmission microscope with detection instrumentation.

itized STEM image and EDX is used to obtain elemental composition information. Phase identification is accomplished by analyzing the electron diffraction patterns. Figure 3 is a TEM micrograph of a Y_2O_3/ZrO_2 powder after calcination at 1000°C. Figure 4 shows the EDX spectrum from the powder. A concentration of 5 wt. pct. Y_2O_3 is obtained from the relative areas under the peaks.

The morphology of two phase polymer blends can be studied by TEM. A microtomed section of a PET/nylon 6 fiber is shown in Figure 5. Phosphotungstic acid was used to stain the nylon and thus enhance the contrast. Nylon inclusions (dark) are seen distributed throughout the PET matrix (light). The size of the inclusions decreases near the edges of the fiber. Phase inversion is seen to occur at various locations.

Bright field, dark field and electron diffraction are used extensively to study the microstructures of ceramic materials, of metallic alloys and of semiconductors. A dark field micrograph of GaAs/GaAlAs interfaces is

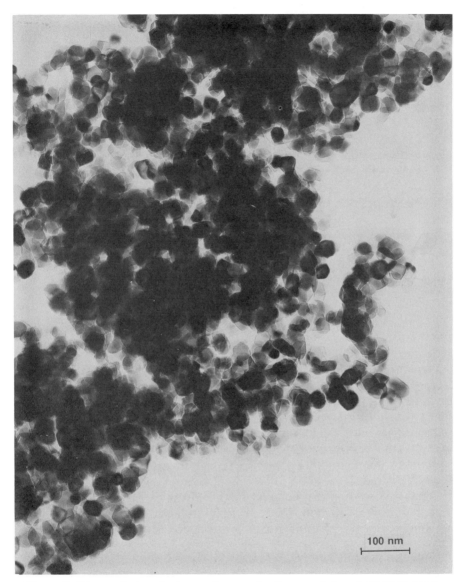

Figure 3. Transmission electron micrograph of co-precipitated 5% Y_2O_3/ZrO_2.

shown in Figure 6. The inset shows the diffraction pattern from the same area. The (200) diffraction spot was used to image the cross-section of the alternating GaAs/GaAlAs layers. In this way the layer thickness and the interface profile can be studied.

The orientation, size and distribution of lamellae in polymer films can also be studied by TEM. Figure 7 shows the orientation of the lamellae in a

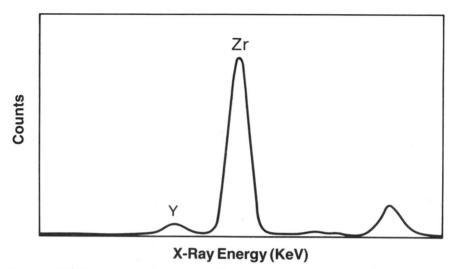

Figure 4. EDX spectrum of sample in Figure 3. From the area under the Y and Zr peaks one obtains the average composition of the sample.

nylon 6 film which was stained with $SnCl_2$ to enhance the electron density of the less ordered regions.

Transmission electron microscopy can also be useful in determining the dispersion of polymer blends. Figure 8 shows a cryomicrotomed section of a nylon/PE graft copolymer. Phosphotungstic acid was used to increase the density of the nylon phase.

Bright field, dark field and electron diffraction techniques are used extensively to study the microstructures of metallic alloys. Figure 9 illustrates the crystallized microstructure of alloy 5006. It consists of Mo/Fe/B precipitates in a Ni-rich matrix.

Replica techniques can be applied to study surfaces of organic and inorganic materials. A carbon replica of PET tire cord showing surface protrusions indicative of crystalline exudates or oligomers is shown in Figure 10.

Limitations

- Sample preparation can be very tedious.
- Some materials, especially polymers, are relatively sensitive to electron beam irradiation resulting in a loss of crystallinity and/or mass.
- Imaging resolution is limited to about 0.2 nm.
- Routine quantitative analysis by EDX is only possible for elements with atomic number > 10 and with a relative accuracy of about 5 to 15%.
- Lighter elements can only be detected with either an EELS or thin window EDX detector and, for the most part, results are only qualitative.

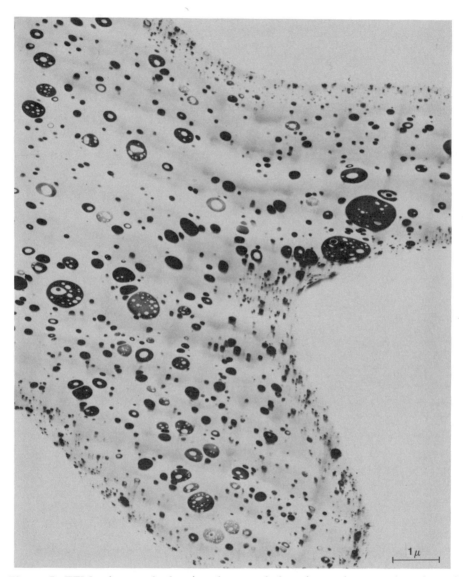

Figure 5. TEM micrograph showing the morphology in a microtomed section of nylon/PET fiber. The nylon inclusions appear dark as a result of the phosphotungstic acid stain used to enhance their contrast.

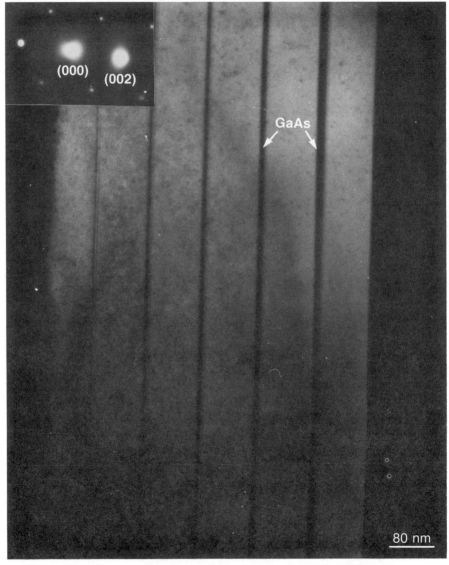

Figure 6. Dark field transmission electron micrograph of alternating layers of GaAs (dark)/GaAlAs (light) epilayers deposited by molecular beam epitaxy onto a GaAs substrate. Inset shows the corresponding diffraction pattern.

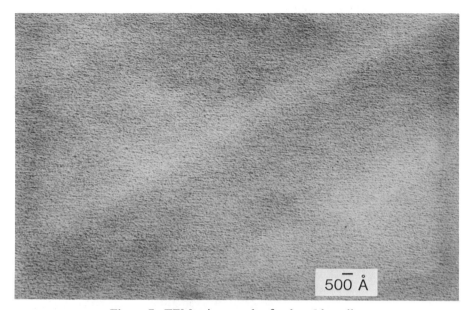

Figure 7. TEM micrograph of nylon 6 lamellae.

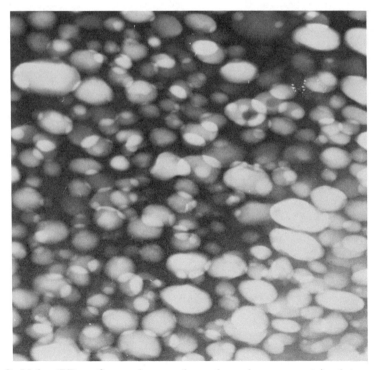

Figure 8. Nylon/PE graft copolymer; the nylon phase was stained to enhance contrast.

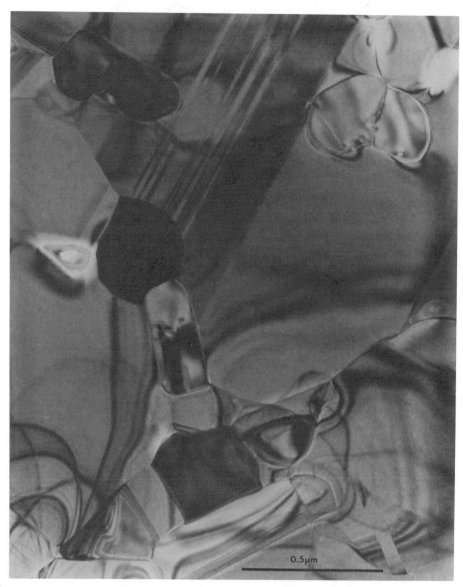

Figure 9. Alloy 5006 microstructure showing Mo/Fe/B precipitates in a Ni-rich matrix.

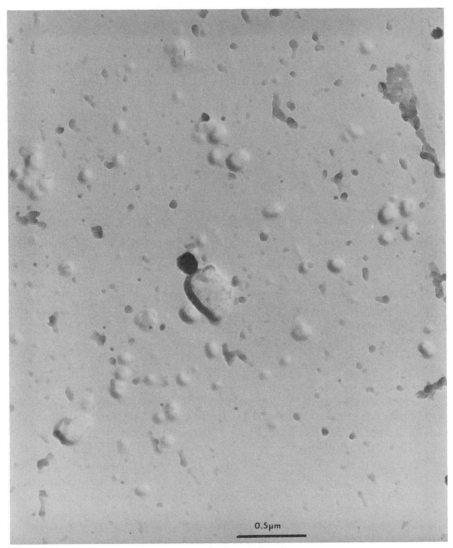

0.5µm

Figure 10. Carbon replica of PET tire cord.

References

(1) Eddington, J. W., Practical Electron Microscopy in Materials Science; Van Nostrand Reinhold: New York, 1976.

(2) Hirsch, P. B.; Howie, A.; Nicholson, R. B.; Pashley, D. W.; Whelan, M. J., Electron Microscopy of Thin Crystals; Butterworths: London, 1965.

(3) Hren, J. J.; Goldstein, J. I.; Joy, D. C., Introduction to Analytical Electron Microscopy; Plenum Press: New York, 1979.

(4) Williams, D. B., Practical Analytical Electron Microscopy in Materials Science; Phillips Electronic Instruments Publishing: Mahwah, NJ, 1984.

AUTOMATIC IMAGE ANALYSIS

Use

Image analysis is used to rapidly and accurately count, measure, and/or classify quantitatively features such as particles, fibers or structural elements.

Sample

Features can be analyzed in both solid or liquid samples as long as the feature of interest can be shown in clear contrast.

Principle

Image analysis can be performed either on a "stand alone" system or on peripheral equipment interfaced to electron beam instrumentation.

The "stand alone" image analyzer is connected directly to an epidiascope or an optical microscope through a video camera. This camera scans the image and generates an analogue signal whose voltage variation corresponds to the intensity variation of the image being scanned. Feature discrimination and selection can now be made on the basis of either grey level values or edges in the image by comparison of the camera output voltage to appropriate reference (threshold) voltages. Finally, after the image is digitized, feature measurement and data reduction are completed.

Peripheral systems interfaced to electron beam instrumentation consist of a digital beam controller, video processor, computer, monitor and peripherals. In such systems, SEM or STEM images are digitized and stored. Feature discrimination is made by grey level differentiation. Figure 1 is a block diagram of an automatic image analyzing system.

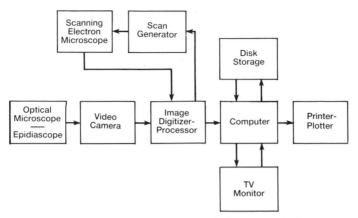

Figure 1. Block diagram of an automatic image analyzing system.

Data obtainable from image analysis include area, perimeter, number of features, longest length, shortest length, equivalent diameter, centroid location and aspect ratio. Data presentation is usually in the form of distribution and cumulative curves for these parameters.

Applications

One of the most frequently used applications of image analysis includes frequency distribution determination (e.g., particle size, length, and aspect ratio distribution of glass fibers or mineral filler in polymer composites and grain size distribution in metals or ceramics).

The best known application for image analysis is its use for particle size analysis, especially when shape factors have to be measured. By using a combination of measuring criteria the size distribution of Wollastonite (an inorganic filler often added to nylon) can be determined based on the longest dimension of the particles, regardless of their orientation. Figure 2 shows a histogram of the frequency distribution of the longest dimension of Wollastonite measured on 2000 particles.

Image analysis has also been used to characterize foamed trilobal fibers. Single filament cross-sections (Figure 3) can be analyzed for "percent void coverage" and modification ratio (circumscribed circle about the cross-section divided by inscribed circle within the cross-section). Modification ratio distribution curves (Figure 4) can be determined and fiber denier calculated.

The image analysis systems interfaced to electron beam instrumentation are supplied with the appropriate software to measure the frequency distribution of particles or phases based on composition in addition to geometric features.

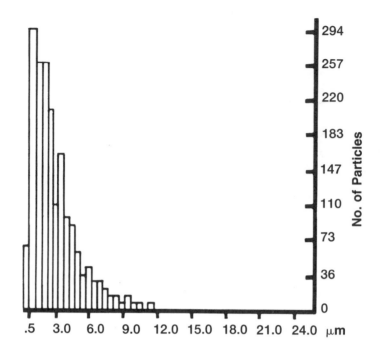

No. of Particles Measured = 2000
Average = 2.91 μm
Range = 0.50 to 65.00 μm
97.1% <10 μm

Figure 2. Longest dimension frequency distribution, Wollastonite.

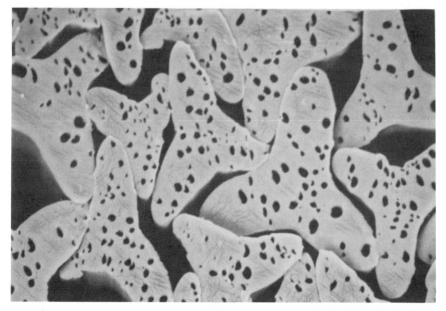

Figure 3. Cross sections of foamed fibers for image analysis.

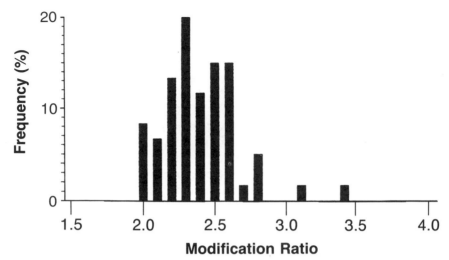

Figure 4. Modification ratio (circumscribed circle divided by inscribed circle) distribution of foamed fibers.

Limitations

Since feature discrimination is based on grey level contrast or edges in the image, it is imperative to choose an imaging mode that allows the feature of interest to be differentiated from the remainder of the field. Additionally, only two dimensional features are measured. Volumetric information of the feature of interest can only be obtained by assuming a specific geometry.

References

(1) Fishchmeister, H., Digital Image Analysis in Quantitative Metallography in Computers in Materials Technology; Pergamon Press: New York, 1981; pp 109–129.

(2) Van der Voort, G. F., Image Analysis, in Metals Handbook 9th Edition, Vol. 10 Materials Characterization; American Society for Metals: Metals Park, 1986; pp 309–322.

MICROSCOPY SPECIMEN PREPARATION TECHNIQUES

Use

Scanning electron microscopy, optical microscopy and transmission electron microscopy require special specimen preparation techniques. These include ultramicrotomy (i.e., thin sectioning), sputtering (i.e., of a thin metallic film), electropolishing, ion milling and various mechanical cutting/polishing techniques.

Sample

In general, the samples should consist of either particles dispersed in a liquid, or solid materials. The dispersions are usually sprayed onto a suitable substrate and coated with a conductive film as described below. Appropriate specimens from solid samples are prepared by applying one or several of the techniques described in the next section.

Principle

a. Cutting/Mechanical Polishing Techniques

Specimen preparation for electron microscopy often includes the use of a diamond wheel cutter. This consists of a thin metal wheel with a diamond abrasive coating. The cutting speed is controlled by adjusting the rotating speed of the wheel. In this way sub-millimeter specimens can be cut with minimum damage.

Mechanical polishing is used to prepare the surface of specimens for microscopical observation. It is also used as a relatively fast way to pre-thin samples. The polishing is done on an automated unit consisting of a rotating wheel and a jig for applying suitable pressure to the sample. Polishing materials including abrasive papers and cloths impregnated with diamond or alumina compounds are used to obtain a lustrous surface.

b. Vacuum Evaporation and Sputtering

Thin conductive films can be deposited by either vacuum evaporation or by sputtering from a target material onto a substrate.

In vacuum evaporation the source material is evaporated by direct heating. The substrate, which is normally at ambient temperature, is rotated during the evaporation procedure to obtain a uniform deposition. Gold and carbon are most commonly used, but other materials such as platinum or palladium are also employed as evaporation sources.

In sputtering, the sample is placed in an inert gas atmosphere (e.g., Ar). A high voltage is applied to cause ionization of the gas. The ions are propelled towards a suitable target, such as a thin gold foil. This causes gold atoms to be knocked off and to be redeposited onto the surface of the sample. In most cases, the films that are deposited are in the range of 10 to 50 nm thick.

c. Ultramicrotomy

Thin sections for optical and for transmission electron microscopy are prepared by microtomy. The sample is mounted on a motor driven stage. Downward movement of the specimen causes it to contact a knife edge. The knife itself is made out of glass or of a diamond single crystal. Often the sample is embedded in an epoxy resin for ease of handling and for stability.

In ultramicrotomy a sensitive mechanism controls the advance of the specimen tip towards the cutting edge of the knife and thus the thickness of the sections. Typically sections with thicknesses in the range of 50 to 100 nm are obtained which are then collected on suitable electron microscope grids or onto glass slides.

If the specimen undergoing preparation is too soft at room temperature, such as polyethylene, cryomicrotomy is used. In this case a special attachment cools the specimen and the knife down to $-160°C$. In this way the distortion of the specimen during sectioning is minimized.

d. Ion Milling

Ion milling is a technique used to thin specimens (e.g., ceramics or metals) for transmission electron microscopy. In this technique a beam of Ar ions impinges on a rotating specimen. As a result of head-on collisions the Ar ions will knock out atoms from the specimen leading to a gradual decrease in thickness.

As soon as perforation occurs, the beam is stopped and the specimen is transferred to the electron microscope for examination. A thin, electron transparent area is usually found near the periphery of the perforation.

Because the rate at which ion milling progresses is only of the order of a few microns per hour the initial thickness of the specimen must be reduced by mechanical polishing to less than 70 μm.

e. Electropolishing

Electropolishing is used to thin metallic specimens which might be affected by the heating effects during ion milling.

During electropolishing, a mechanically thinned specimen is immersed in an acid electrolyte. A potential is applied between the specimen and two parallel plate cathodes positioned on opposite sides of the specimen. As a result of the electrochemical reaction, specimen dissolution occurs. When a small perforation forms the reaction is stopped; the specimen is then rinsed and examined in the microscope.

The temperature of the electrolyte, the voltage and the concentration of the electrolyte are variables which will affect the uniformity with which the specimen dissolves.

Limitations

Often specimen preparation is more an art than a science. Patience is a key ingredient in the preparation of specimens for microscopical analysis.

In general, mechanical polishing will be sufficient if specimens are to be examined by either optical or scanning electron microscopy. Mechanical polishing, however, must be followed by either electropolishing or ion milling if transmission electron microscopy is to be performed. Electropolishing is only applicable to metallic specimens but thinning rates are considerably faster than in ion milling. On the other hand, finding the right combination of electrolyte, temperature and voltage for a particular sample can be a tedious exercise.

Ultramicrotomy is not suitable for metallic or ceramic samples unless these are in the form of very fine powders. Diamond knives are recommended over glass knives but the former are considerably more expensive.

Thin films produced by vacuum evaporation tend to have a finer structure than those produced by sputtering and they are better suited for TEM applications where higher resolutions are required. Conductive films for SEM applications can be produced in a shorter time by sputtering.

References

(1) Kay, D., Techniques for Electron Microscopy; Blackwell Scientific Publications: Oxford, Great Britain, 1967.

(2) Hayat, M. A., Principles and Techniques of Electron Microscopy; Van Nostrand Reinhold Co.; New York, Vol. I, 1970; pp 183–237.

(3) Reed-Hill, R. E., Physical Metallurgy Principles: Van Nostrand Co.: New York, 1973; pp 2–5.

(4) Thomas, G., Transmission Electron Microscopy of Metals; John Wiley and Sons, Inc.; New York, 1966; pp 133–182.

CHAPTER 8

Surface Analysis

Anthony J. Signorelli, Edgar A. Leone, Roland L. Chin

X-RAY PHOTOELECTRON SPECTROSCOPY

Use

X-ray photoelectron spectroscopy (XPS) or electron spectroscopy for chemical analysis (ESCA), is a surface sensitive spectroscopic tool that provides information about the composition and structure of the outermost surface layers of a solid.

Sample

Any solid material can be studied using XPS. All elements, with the exception of H and He, can be detected.

Principle

When a solid is exposed to a flux of x-ray photons of known energy, photoelectrons are emitted from the solid. These photoelectrons originate from discrete electronic energy levels associated with those atoms in the analysis volume. The energy of the emitted photoelectrons is given by:

$$E_k = h\nu - E_B - \phi$$

where $h\nu$ is the characteristic photon energy of the excitation source, E_k and E_B are the measured photoelectron kinetic energy and binding energy of a specific core or valence level electron, respectively, and ϕ is a parameter which depends on the spectrometer and the sample being analyzed. Since ionization may occur in any shell for a particular atom (with varying probability) the spectrum for that element is usually comprised of a series of peaks corresponding to electron emission from the different shells. This allows for unambiguous elemental identification since the energy separation and relative intensities of the peaks for a given element are well known. Additionally, ionization for p, d or f levels leads to doublet structure in the spectrum as a result of spin-orbit interactions (i.e. generation of two different j states). In many cases, the spin-orbit separation energy between the doublet peaks is characteristic of the element in a particular oxidation state.

Of greatest diagnostic utility is the dependence of E_B on the oxidation state and/or local electronic environment about the particular atom being probed. These variations in E_B are commonly referred to as chemical shifts and are due to a variety of factors:

1. Compounds and ions containing nonequivalent atoms of the same element such as N_2O, $S_2O_3^{-2}$ and $CF_3COOCH_2CH_3$
2. Solid materials having atoms of different oxidation state present in nonequivalent lattice sites. The best known examples are the mixed valency metal oxides which include such compounds as Fe_3O_4 ($Fe^{2+} - Fe^{3+}$), KCr_3O_8 ($Cr^{3+} - Cr^{6+}$) and Pb_3O_4 ($Pb^{2+} - Pb^{4+}$).
3. Variations in the core level binding energy of an element when it is present in different compounds (eg. $Mo°$, MoS_2 and MoO_3).

In addition to the ability to differentiate nonequivalent atoms, the relative signal intensities provide a direct measure of the stoichiometry.

Due to the complex nature of the x-ray photoemission process, a number of so-called final state effects can be observed which prove to be of diagnostic value. These may include the emission of Auger electrons which provide information concerning the elemental and chemical composition of a sample; lifetime broadening effects which are studied in order to understand the fundamental photoemission event; multiplet effects which are influenced by the unpaired electron spin density of a material; and shakeup phenomena which in many cases provide both chemical and structural information.

Finally, electrons are characterized by very short mean free paths within a solid which is energy dependent. The most commonly used x-rays for excitation are the Mg K_α and Al K_α at energies of 1253.6 eV and 1486.6 eV, respectively. Typical sampling depths would range between 0.5 and 5 nm.

However, x-ray sources of higher energy are often employed in order to increase the probing depth. Due to the extreme surface sensitivity of XPS, contamination of the sample must be kept to a minimum. This requires the use of ultra-high vacuum equipment (1×10^{-9} torr) to maintain surface cleanliness. A schematic diagram of a typical XPS spectrometer is given in Figure 1.

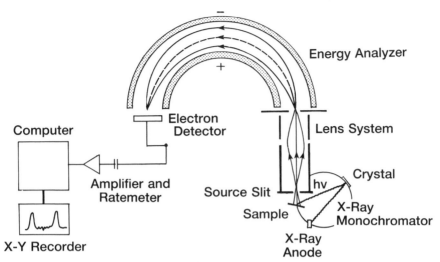

Figure 1. Schematic representation of an XPS spectrometer employing a hemispherical electron energy analyzer and monochromatic x-ray source.

Applications

XPS is applicable to the study of heterogeneous catalysts, polymers, problems associated with adhesion or wettability, ceramics, corrosion products, metals, alloys, and semiconductors. In addition to providing information related to electronic structure, the technique allows one to determine the average surface composition of a solid within an analysis area that is typically several mm^2. By employing selective aperturing in the electron optics of the transfer lens and analyzer, analysis of areas ranging in size from 100 to 150 microns can be obtained.

The extreme surface sensitivity of the technique is exemplified by the spectra shown in Figure 2. It was suspected that during the high temperature processing of $LiNbO_3$ substrates, Li out-diffused to the surface. A comparison of the Nb4s/Li1s ratio before (Figure 2a) and after (Figure 2b) thermal treatment clearly shows an increase in Li concentration at the surface.

An example where XPS was used to evaluate the cause of poor adhesion is illustrated by the analysis of a laminate film used as a packaging material. The laminate exhibited areas characterized by poor and good sealability. The resultant XPS spectra for poor and good seal areas revealed that the poor seal area was contaminated by a thin aluminum oxide layer (Figure 3).

An example of adhesive failure resulting from thermal cycling involved the delamination of a fluoropolymer film from an epoxy impregnated back-

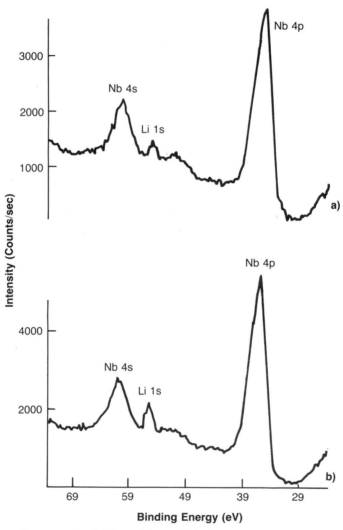

Figure 2. XPS spectra for lithium niobate (a) before and (b) after thermal treatment.

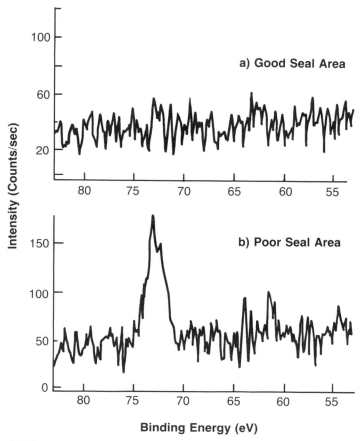

Figure 3. XPS spectra showing (a) a good seal area and (b) the presence of a contamination layer of aluminum oxide on the surface of a film having poor sealability.

ing board composite. Examining the N and F core levels from areas that did not delaminate (Figure 4a) and corresponding core levels from areas that underwent delamination (Figure 4b) revealed that an amine curing agent had attacked the polymer resulting in the formation of a weak interfacial layer that had a composition consistent with ammonium hydrofluoride.

The wettability of Pb-In-Ag foils and a particular substrate were found to change as a function of the age of the foil. Typical Pb4f core level spectra are shown in Figure 5. Note that the surface of the aged foil has a thicker lead oxide layer when compared to a fresh foil. In addition, the $In3d_{5/2}$ to Pb4f ratios suggest that In migrates to the surface with time. The changes in the surface composition with time were found to correlate with the changes in wettability.

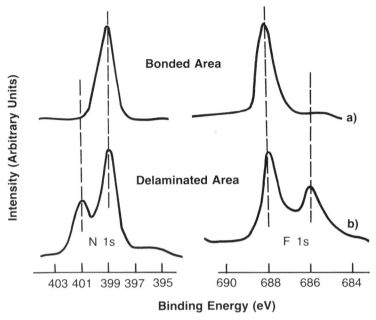

Figure 4. The N1s and F1s spectra associated with (a) bonded and (b) delaminated areas.

An example of the application of XPS to catalyst studies is shown in Figure 6. A silica supported Cu catalyst was examined in order to determine changes in the Cu oxidation state as a function of temperature and H_2 treatment. Representative Cu spectra are shown in the insert. This *in situ* experiment indicates that at elevated temperatures, with or without H_2, there was a gradual reduction of Cu^{+2} to Cu^{+1}. Subsequent studies using oxhydrochlorination conditions established that an active catalyst is defined by a specific Cu^{+1}/Cu^{+2} ratio. These studies were carried out using an environmental chamber which is interfaced to the spectrometer.

Figure 7 reproduces the aluminum 2p spectra (from top-to-bottom) of Al foil, Al foil after compression molding with hexafluoro isobutylene-vinylidene fluoride copolymer, and AlF_3. Note that a significant portion of the Al substrate has been fluorinated. The data suggests that during compression molding, HF is liberated presumably due to degradation of $(CH_2CF_2)n$ blocks. This fluoride-attacked layer is probably only 0.5 to 1 nm in thickness, for Al_2O_3 is still detectable.

Treatment of polymers in a plasma discharge is a common technique for enhancing adhesion properties. Figure 8 shows the change in the C and O spectra for poly(ethylene terephthalate) (PET) that has been corona treated

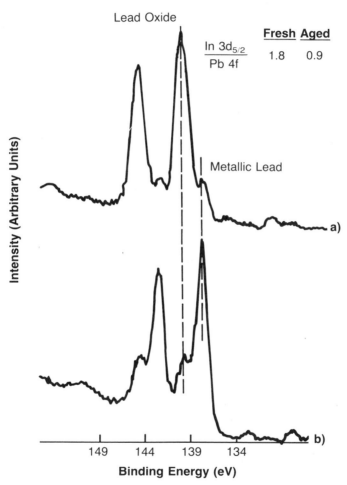

Figure 5. Typical Pb 4f XPS spectra for (a) aged and (b) fresh metal foil.

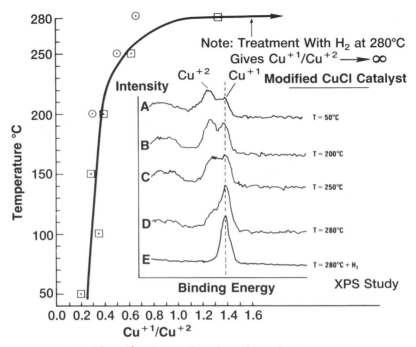

Figure 6. Plot of Cu^{+1}/Cu^{+2} ratio as a function of reaction temperature; corresponding Cu 2p XPS core level spectra shown in insert.

(lower curves). Note the dramatic increase in O intensity and the increased concentration of

$$\underset{R\text{-}C\text{-}O, \; R\text{-}C\text{-}O, \; \text{and} \; R\text{-}C = O}{\overset{\displaystyle O}{\overset{\displaystyle \|}{}}}$$

species on the surface. In addition, functionalization techniques have been developed which allow one to determine the relative concentrations of carboxyl and hydroxyl groups on the surface. This technique is of interest in the field of polymer adhesion.

Limitations

Due to the focusing limitations of x-rays, XPS is usually considered a broad beam technique with an analysis area of about a square centimeter. In small area mode, the analysis diameter can be reduced to about 150 μm. However, in this mode the analysis time is considerably increased due to a severe loss in intensity. In addition, certain materials are sensitive to surface

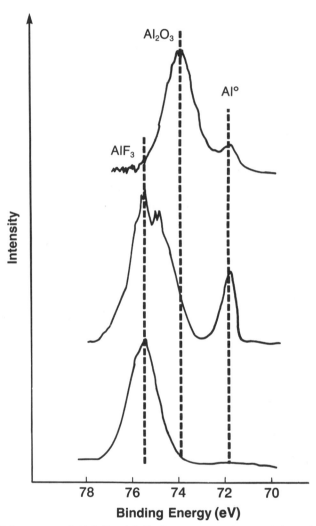

Figure 7. XPS spectra of Al foil; Al foil after compression molding with a fluorine containing polymer; AlF₃.

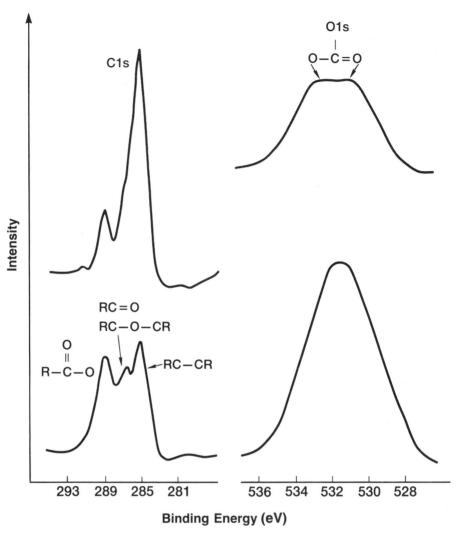

Figure 8. XPS spectra of PET (top) and corona treated PET (bottom).

photoreduction and ion beam damage effects. It is difficult to analyze liquids and gases due to the UHV environment.

References

(1) Briggs, D. ed.; Handbook of X-Ray and Ultraviolet Photoelectron Spectroscopy; Heyden Press: London, 1972.

(2) Fadley, C. S., Electron Spectroscopy: Theory, Techniques and Applications; Vol. 2, Brundle, C. R.; Baker, A. D. eds.; Academic Press: London, 1978.

(3) Feuerbacher, B., Photoemission and the Electronic Properties of Surfaces; J. Wiley and Sons: New York, 1978.

(4) Riggs, W. M.; Parker, M. J., Methods of Surface Analysis; Czanderna, A. W. ed.; Elsevier Publishing: New York, 1975.

SCANNING AUGER MICROSCOPY

Use

Scanning Auger microscopy (SAM), with a probing depth of 1-5 nm, is used principally for identifying chemical constituents at the surface and at interfaces of materials. Spot analyses are on the order of 100 nm, or alternatively, the electron beam can be rastered over much larger areas (0.5 mm^2) to obtain average surface compositions. In addition to the secondary electron images which provide topographic information, spectra, profiles, element specific line scans and maps can be obtained and stored for subsequent computer processing.

Sample

The sample size for Auger analysis can range from a few millimeters to about a square centimeter. Conducting materials such as metals and semiconductors can be analyzed directly. Insulating samples (glass, polymers, etc.) are much more difficult to analyze due to charging effects. Optimizing electron beam conditions and operating in a glancing angle mode provide some flexibility in analyzing non-conducting specimens.

Principle

In Auger electron spectroscopy (AES) the excitation source is a finely focused electron beam which impinges on the sample surface. The two interactions of interest in SAM and AES are the generation of Auger electrons via the Auger process and the production of secondary electrons providing topographic information. The Auger process is initiated by electrons in the primary beam which cause the ejection of core level electrons in the atoms of the sample. This is shown schematically in the energy level diagram of Figure 1. Once a core hole is created, an electron from a higher energy level can then fall into the core vacancy. The energy of this transition can be released in one of two ways: 1) by the release of characteristic x-rays; or 2)

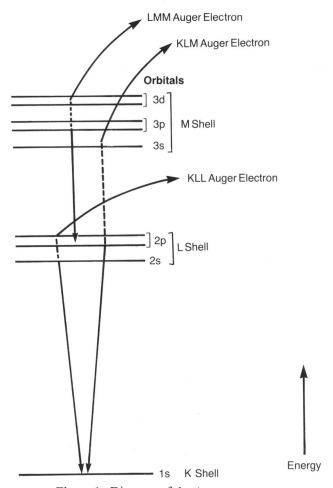

Figure 1. Diagram of the Auger process.

by the ejection of a second (Auger) electron. The process whereby a K shell ionization occurs followed by an L shell electronic transition giving rise to an L shell ejected Auger electron is termed a KLL transition. Similarly, other transitions are possible such as KLM and LMM as indicated in Figure 1. The Auger electron then escapes into the vacuum and enters a suitable detection system resulting in a display of the number of electrons N(E) vs. their kinetic energy (KE). Because of the relatively high secondary electron background, Auger spectra are usually displayed in the derivative mode, i.e., dN(E) vs. KE.

In general, AES (or SAM) provides identification of the elemental composition of a sample. For a limited number of cases, chemical speciation is derivable. Unlike XPS, the sampling depth is independent of the energy of the excitation source, since the emission of the Auger electron is not directly coupled to the primary ionization event. However, for the most commonly used XPS sources, the sampling depths of the two techniques are comparable. The primary advantage of AES as a surface analytical tool is the small spot size of the electron beam probe which allows for analysis of sample features which are beyond the spatial resolution of other commonly employed surface techniques. Because of its inherently shallow sampling depth and high spatial resolution, AES in combination with ion sputtering may be used for depth profiling in order to determine the composition of a sample as a function of depth below the surface. A schematic diagram of a typical AES spectrometer is given in Figure 2.

Applications

An analysis which takes advantage of the high spatial resolution provided by SAM, involves the identification of the cause for tarnishing of a Au/Sn alloy which was bonded to a Au plated iron containing substrate. An elemental survey scan (Figure 3a) showed that C, O and Fe were the only elements present at the surface. The bottom spectrum (Figure 3b) reveals the Au and Sn signals of the alloy after removal of the iron containing layer. It was determined via Auger depth profiling analysis that the tarnish resulted from the surface segregation of iron oxide which was ~100 nm in thickness. Further analysis revealed that the Au plating had partially alloyed with the molten Au/Sn alloy during processing. This facilitated the rapid diffusion and oxidation of Fe from the underlying substrate.

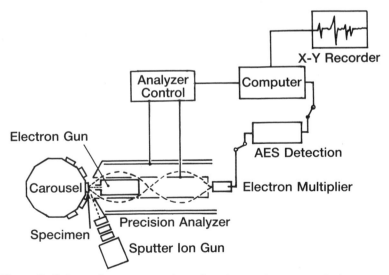

Figure 2. Schematic representation of an Auger electron analysis system.

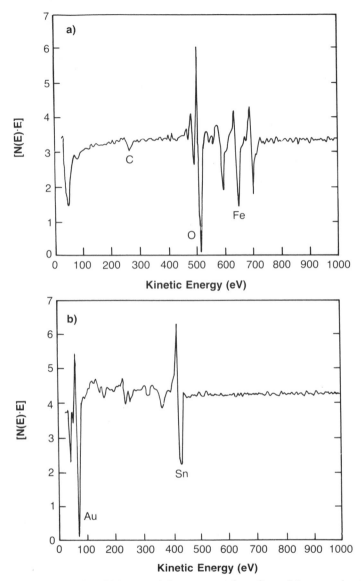

Figure 3. Auger analysis of blue tarnish on a metal preform (a) as received and (b) after Ar ion sputtering.

Depth profiling is an important mode of Auger analysis which provides estimates of the relative thickness of surface and/or interfacial layers. A profile is obtained by generating spectra of the elements of interest while simultaneously removing a thin layer of material via ion sputtering. This procedure is repeated many times resulting in a display of elemental intensity vs.

sputter time. An example of how this can be used to determine relative changes in elemental composition as a function of processing conditions is shown in Figure 4. The uppermost display is a profile of the wheel side of an "as cast" METGLAS® ribbon. A typical oxide thickness is on the order of 5-7 nm. The bottom profile is from an annealed ribbon showing an

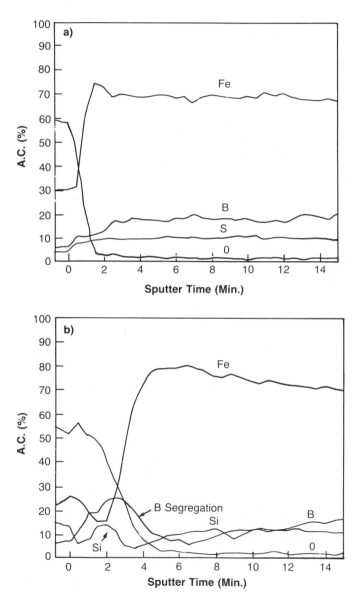

Figure 4. Depth profile analysis of METGLAS® ribbons (a) as cast and (b) after thermal aging.

increase in the thickness of the surface oxide and additional segregation of B and Si into the surface oxide layer.

Auger electron mapping is a very useful way of observing the lateral elemental distribution of very thin surface residues. An example of the use of the technique for solving contamination problems in semiconductors is given as follows. During the fabrication of semiconductor wafers, a dark

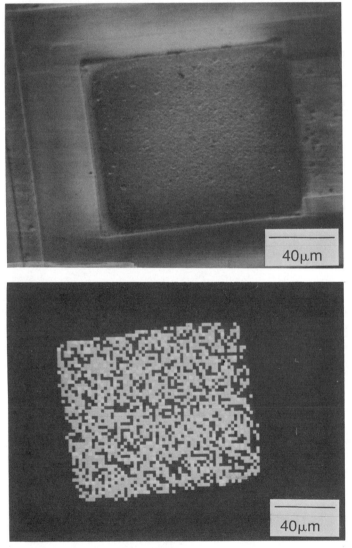

Figure 5. (a) SEM image of a bonding pad and (b) fluorine Auger map of the same area.

staining problem was observed optically on some of the Al bonding pads used to bond wire leads to the internal circuitry. SAM analysis of these pads showed that F, S and C were present along with the expected Al and O signals. This indicated that a corrosive agent containing F and S may have been present during the processing of the wafer. The SEM image and a F Auger map of one of the pads are shown in Figure 5. The computer enhanced image reveals that the F was present only in the vicinity of the bonding pads and not as a surface contaminant over the rest of the wafer.

Some further examples of the applicability of AES to materials characterization are given in Figures 6 and 7. (Examples courtesy Physical Electronics Industries, Inc.)

A junction region of an alloyed Al-Si contact was depth-profiled in order to ascertain the cause of the high contact resistance measured for a particular device. The compositional profile (see Figure 6) indicates that at the Al-Si interface there is a definite oxygen contamination.

Small amounts of metallic impurities on the surface of metallic substrates can give rise to poor thermo-compression bonding when attaching wire leads. The Auger data presented in Figure 7 clearly indicate that the device exhibiting poor bonding characteristics has a high Ag surface (0.5 nm thick)

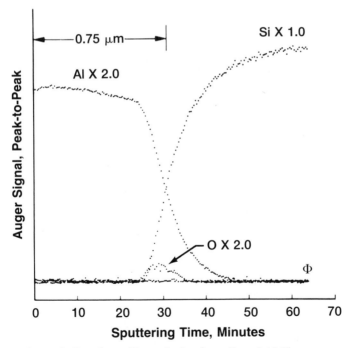

Figure 6. Depth profile analysis of an alloyed Al-Si contact.

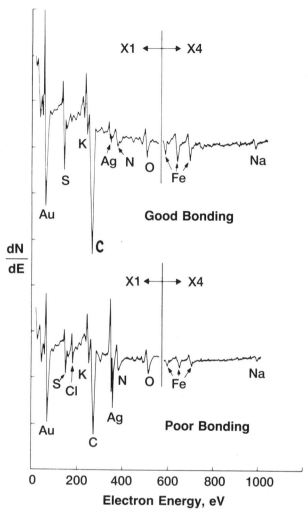

Figure 7. AES spectra of bonding pads exhibiting good bonding (top) and poor bonding (bottom).

concentration. An analysis of this sort would not be possible by bulk analytical techniques.

Limitations

As mentioned previously, sample charging makes it extremely difficult to analyze insulating materials. Although the spatial resolution is excellent, as the analysis size decreases, electron beam damage can alter the surface composition. Due to the nature of the Auger process, the peaks are generally rather broad so that spectral over-lapping can be a problem.

References

(1) Carlson, T. A., Photoelectron and Auger Spectroscopy; Plenum Press: New York, 1975.

(2) Fuggle, J. C., Electron Spectroscopy: Theory, Techniques and Applications, Vol. 4, Brundle, C. R.; Baker, A. D., eds.; Academic Press: London, 1981; 85.

(3) Joshi, A.; Davis, L. E.; Palmberg, P. W., Methods of Surface Analysis; Czanderna, A. W., ed.; Elsevier Publishing: New York, 1975; 159.

(4) Springer, R. W.; Haas, T. W.; Grant, J. T., Quantitative Surface Analysis of Materials, ASTM STP 634; American Society for Testing and Materials: Philadelphia, 1978; 64.

Acknowledgment

Figure 2 is reprinted from: Kirk-Othmer, Encyclopedia of Chemical Technology; 3rd ed. John Wiley and Sons, Inc.: New York, copyright© 1978; 673, with permission of John Wiley and Sons, Inc. and through courtesy of Physical Electronics.

Figures 6 and 7 are reprinted from: PHI Application Note, 1979; with permission of the Perkin-Elmer Corporation.

SECONDARY ION MASS SPECTROMETRY

Use

Secondary ion mass spectrometry (SIMS) is capable of providing surface and elemental depth concentration profile analysis on areas ranging from several square millimeters down to the submicron level. When the SIMS technique is used at the submicron level, it can provide ion induced secondary electron images (surface topography) as well as element specific maps.

Sample

Any solid material can be studied. Typically, samples not larger than 1 cm^2 and as small as a few microns can be analyzed.

Principle

In the SIMS experiment, a primary beam of energetic ions is directed at a specimen, resulting in the ejection of secondary ions from the sample surface. The process is schematically represented in Figure 1. The ejected fragments have relatively low energy (5–50 eV), and consist of positively and negatively charged ions, neutral atoms and molecular fragments. Classical SIMS uses inert gas ions such as He^+, Ar^+, Xe^+, etc., as the primary ion source. The inert gases are introduced into a chamber where ionization occurs via electron impact. The positively charged ions are extracted from the ion source, accelerated to a kinetic energy of several keV and focused on the sample.

The probe diameters can range anywhere from 10 μm to 1 mm. Using noble gas ions, it is difficult to obtain probe diameters < 5 μm.

With the recent advent of liquid metal ion (LMI) sources, it is possible to obtain positively charged ion beams with diameters < 50 nm. Generally, metals with relatively low melting points, such as Cs and Ga are used as the ion source. The liquid metal is drawn over the tip of a fine metal needle via capillary action. By means of a large potential field applied between the needle and an extractor plate the metal is distorted to produce a cone. The positive metal ions are extracted from this cone (essentially a point source) which can be focused and rastered. As a result, a secondary electron image can be obtained in much the same fashion as in SEM. However, the contrast mechanisms are slightly different since the exciting beam is composed of positive ions rather than electrons.

There are basically three different types of secondary ion detection systems presently employed in SIMS analyzers. Quadrupole mass spectrometers have a mass resolution of about one atomic mass unit (amu), a mass range up to 800 amu and have a collection efficiency of 1–5%. A simplified schematic diagram of a typical quadrupole based LMIG-SIMS system is shown in Figure 2. The best mass resolution can be obtained with a mag-

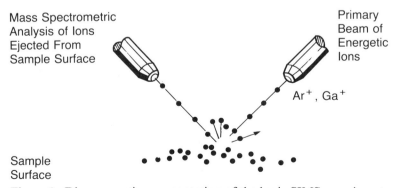

Figure 1. Diagrammatic representation of the basic SIMS experiment.

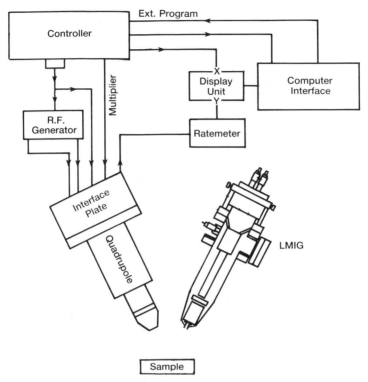

Figure 2. Simplified block diagram of a quadrupole based LMIG-SIMS system.

netic sector spectrometer, which typically has a mass resolution of 25,000 or better. Time-of flight (TOF) is a third type of SIMS spectrometer, which has recently gained some attention. In this system, an LMI beam is pulsed with a variable pulse width of 10 ns. The secondary ions are extracted from the sample surface, mass separated by a flight tube and detected quasi-simultaneously using a channel plate detector. There is no fixed upper limit of mass detection, since the maximum mass depends upon the square of the flight time, and the pulse frequency can be decreased to accommodate larger masses (5000 amu is not unreasonable). The collection efficiency of TOF-SIMS is nearly 100%. It is interesting to compare the relative ion doses required to obtain equal signal intensities for the various SIMS analyzers. On a 300 amu scale, a magnetic sector spectrometer requires an ion dose 10^4 times greater than a TOF system. Quadrupoles are even higher, requiring doses one to three orders of magnitude greater than the magnetic sector systems.

SIMS analysis can be classified into three categories depending upon the incident ion dosage. Static SIMS requires that the ion dosage per unit area be kept low, typically $< 10^{13}$ cm^{-2}, so that undamaged surface molecular fragments can be detected. Low energy (1-2 keV) noble gas ions or FAB sources are commonly used. In a dynamic SIMS experiment, the ion dosage, and thus the sputter rate, is correspondingly higher. This results in greater fragmentation, and hence a shift to lower molecular weight species, even to atomic or diatomic components. When used in the imaging mode, the LMI source is rastered over the sample surface and the output intensity of a specific ion can be displayed on a CRT screen, or stored in a computer. Because of the high spatial resolution of the LMI source, it is possible to obtain images showing the sub-micron distribution of elements in the surface layers. Imaging is a dynamic process, and as such, changes in surface composition can be monitored with time.

Another factor that is critical to SIMS analysis, is the requirement for charge neutralization when analyzing insulating materials. In dynamic and imaging studies where the incident ion current is large, a positive charge builds up on the sample surface, preventing the secondary ions from escaping. Charge neutralization of this positive charge is provided by carefully adjusting the intensity and focus of a low energy (500 eV) electron source. The process of charge neutralization, the efficiency of the detection system and the incident beam current are intimately related. As the detection efficiency increases, less primary beam current is required. This results in smaller probe diameters (i.e., higher spatial resolution) and reduced need for charge neutralization.

The characteristic which distinguishes SIMS from the other forms of surface spectroscopy is an overwhelming advantage in detection sensitivity. For example, whereas the detection limits for XPS and AES are on the order of parts per thousand, limits in the ppm and ppb range can be realistically achieved by SIMS. Therefore, applications which involve trace analysis such as dopant distribution in semiconductors are possible by SIMS. The high sensitivity of SIMS is not altogether surprising since other mass spectrometry (MS) techniques have comparable sensitivities. However, unlike the other MS techniques (eg., gas phase), SIMS is not readily amenable to quantitative analysis. This is due to the complex nature of the ionization process of surface atoms and molecules which is highly dependent upon the sample matrix. Thus, for a particular element, detection limits may vary several orders of magnitude since the ion yield is greatly influenced by the matrix. This problem can be circumvented by the use of standards whose matrix is identical to that of the unknown, although only in a limited number of cases is this approach feasible.

Applications

A wide variety of solid samples can be analyzed using SIMS. These include both conductors and nonconductors alike. For example, SIMS can be used to study metals, ceramics, fracture surfaces, catalysts, semiconductor materials and finished devices, optical fibers and substrates, polymers and polymer finishes. As examples of the use of SIMS, some recent applications of the technique are described.

In the dynamic mode, SIMS can be used to provide information about the distribution of elements as a function of depth. In this regard, SIMS provides data that is not unlike Auger depth profiling. However, SIMS profiles can be performed on non-conductors as well as conducting specimens. In addition, relative to Auger, SIMS has a much higher surface sensitivity. Figure 3a is a SIMS depth profile of an "as-cast" METGLAS® ribbon. The profile was obtained using an inert ion beam as the probe and without O_2 dosing (a technique that enhances detection sensitivity). After exposing the ribbon to high temperature in air, the SIMS depth profile (Figure 3b) clearly shows that the oxide depth has increased by nearly a factor of three. In addition, note that the Si and B have diffused to the surface.

Optical modulators and directional couplers can be fabricated from Ti in-diffused $LiNbO_3$ strip waveguides. The quality of the fabricated device, is, to a certain extent, dependent on the lateral and vertical distribution of the Ti in the in-diffused channels. SIMS, using a Ga^+ ion probe, is ideally suited for determining the lateral diffusion width as well as the diffusion depth of the Ti. An example of the former type of profile is shown in Figure 4. A schematic representation of the patterned Ti in-diffused $LiNbO_3$ substrate is shown in the insert. The SIMS map and the accompanying SIMS linescan provide information about the X-Y or lateral Ti concentration.

As indicated above, a metal ion probe is well suited for the study of polymers and polymer finishes. Figure 5 is a $^{19}F^-$ SIMS map showing the dispersion of a fluorine containing additive that imparts soil repellency characteristics to nylon carpet fibers. Note from the SIMS map that, for the most part, the additive is confined to the surface of the trilobal fiber. Furthermore, the additive dispersion appears to be fairly uniform.

In another example, electrical measurements on a 500 nm layer of pure SiO_2 deposited on an integrated circuit chip had shown that the film was contaminated with a high concentration of mobile ions. AES was unable to detect the contaminants, but SIMS studies using an Argon ion source indicated that poorly performing devices were characterized by a high concentration of Na, Li, K, and B. These contaminants were linked with the method for packaging the integrated circuit chips. The SIMS line scans for the control specimen and contaminated specimen are given in Figures 6a and 6b.

Limitations

Although SIMS has potentially the highest surface sensitivity, it is the least quantitative of the surface techniques. When employing quadrupole based detection systems, the mass resolution is on the order of 0.5–1 AMU

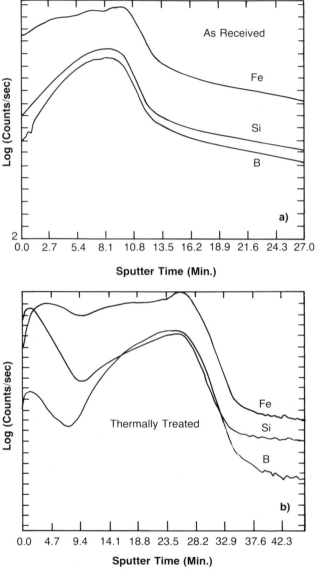

Figure 3. SIMS depth profile for (a) as received and (b) thermally aged METGLAS® ribbon.

which is inadequate in applications where high mass resolution is required. In dynamic or imaging modes, considerable amounts of material are sputtered from the sample surface so that the surface composition can be changing with time. Also, when analyzing insulating materials, electron beam

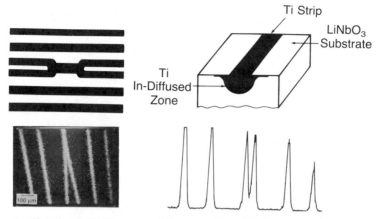

Figure 4. Titanium SIMS map and linescan for a fabricated electro-optic device.

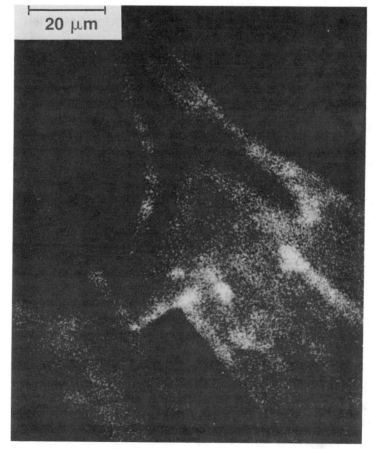

Figure 5. Fluorine ion SIMS map for an ANSO IV fiber.

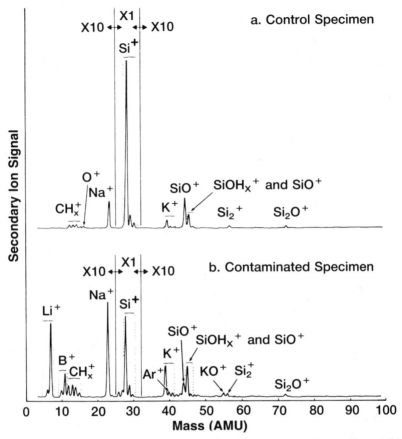

Figure 6. SIMS spectra of deposited S_1O_2 layers for (a) control sample and (b) contaminated sample.

charge neutralization is sometimes required. This can affect relative intensities and lead to surface desorption of sensitive elements.

References

(1) Benninghoven, A.; Giber, J.; Laslo, J.; Riedel, M.; Werner, H. W., eds.; Secondary Ion Mass Spectrometry: Springer-Verlag: New York, 1982.

(2) Colton, R., Nuclear Instru. and Meth. in Phys. Res., 218, 1983; 276.

(3) Liebl, H., Advances in Mass Spectrometry; Heyden Press: London, 1978; 751.

(4) Werner, W.; Boudewign, P. R., Vacuum, 34(1–2), 1984, 83.

(5) Wittmaack, K., Surface Science, 89, 1979; 668.

Acknowledgment

Figure 2 is reprinted in part from: SQ300 SIMS/RGA Quadrupole Mass Spectrometer Booklet; through courtesy of V.G. Instruments Inc.

Figure 6 is reprinted from: PHI Application Note, 1979; with permission of the Perkin-Elmer Corporation.

ULTRAVIOLET AND BREMSSTRAHLUNG ISOCHROMAT SPECTROSCOPY

Use

Ultraviolet photoelectron spectroscopy (UPS) and Bremsstrahlung isochromat spectroscopy (BIS) are branches of electron spectroscopy that can provide information about the occupied and unoccupied density of states in metals and alloys. In addition, UPS can be used to study molecular orbital characteristics, as well as the nature of the interaction between an adsorbate and substrate.

Sample

For UPS, any solid sample can be used. For BIS, conducting samples such as metals or alloys are preferred.

Principle

Phenomenologically, the basic underlying principles that apply to XPS also apply to UPS. The important difference between XPS and UPS is the energy of the excitation probe. In UPS, the photon flux is not provided by an x-ray source ($h\nu > 1000$ eV) but by a dc-induced discharge in flowing helium. As a result of the discharge, photons of energy 21.2 eV (HeI resonance line) and 40.8 eV (HeII resonance line) are produced. These photon energies are ideally suited for the study of the occupied density of states in metals, alloys and semiconductors (Figure 1b). In addition, high resolution spectra can be obtained for the valence band(s) of other solid materials such as inorganic compounds and polymers.

The valence band spectra provide information about molecular orbital parentage or make-up. Furthermore, the electronic nature of adsorbate-substrate interactions can be determined. For example, if E_g^i represents the measured gas phase binding energy associated with the ith molecular orbital electron and $\bar{E}_{ads}^i$ is that same molecular orbital's electron binding energy in the adsorbed state, then

$$E_g^i - E_{ads}^i = \phi + \Delta E_B^i + \Delta E_R^i$$

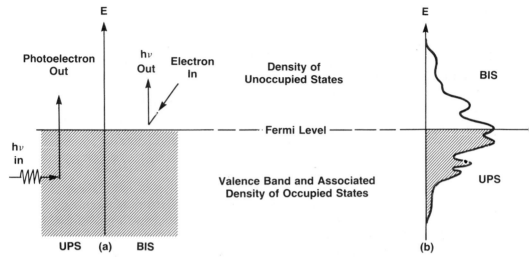

Figure 1. Schematic representation of the UPS and BIS experiment.

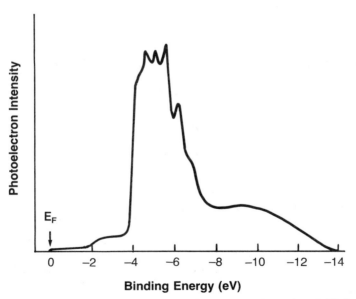

Figure 2. The uncorrected UPS spectrum for the Ag valence band. He 1 source.

Thus, from a knowledge of the substrate (solid) work function ϕ and a measurement of the change in relaxation, ΔE_R^i, one can determine the chemical shift, ΔE_B^i, induced by bonding differences.

BIS involves the inverse process, i.e., an electron of kinetic energy E_K incident on a solid undergoes a radiative transition (Bremsstrahlung) as it occupies a previously empty state above the Fermi level (Figure 1a). This process can be thought of as the inverse of photoemission, i.e., in normal photoemission an incoming photon brings about the transition of an electron from a normally occupied state to a free electron state; in BIS a photon is emitted as an electron makes a transition from a free electron-like state to a state above the Fermi level. By varying the incoming electron kinetic energy and detecting a fixed photon energy (isochromat) one can map out the density of unoccupied states or those electronic states that reside above the Fermi level (Figure 1b).

Applications

UPS and BIS can be used to map the density of occupied and unoccupied states, respectively. This information is of relevance when dealing with magnetic and electrical/electronic properties of metals, semiconductors, and conducting polymers. Both techniques can provide information about oxidation state and the chemisorption process.

As examples of how UPS and BIS can be used to provide insight into the nature of the valence band and the density and symmetry of states above the Fermi level the uncorrected UPS spectrum for Ag (Figure 2) and the BIS spectrum for Ni (Figure 3) have been reproduced. For Ag the most intense features are due to the 4d bands. Onset for d-photoemission starts at approximately -3.9 eV (or 3.9 eV below the Fermi level) and the d-band width is approximately 3.5 eV. The weak emission from E_F to -1.9 eV is due to bands of sp symmetry. The BIS spectrum for Ni indicates a maximum in the unoccupied density of states at ~ 0.3 eV above the Fermi edge. These are states associated with d-symmetry. Signal averaging the data could possibly reveal other peaks between 0.3 eV and 6 eV above E_F which may be related to states of s and p symmetry.

Limitations

Radiation damage to the sample must be considered when performing analysis by UPS or BIS. This situation is particularly true for organic samples. The use of an electron beam in BIS may result in desorption of certain

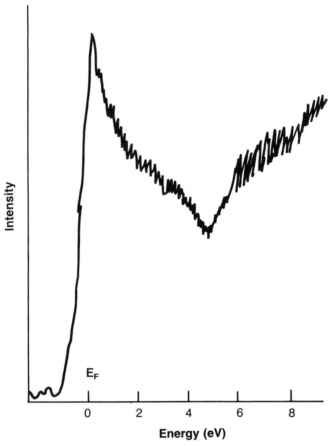

Figure 3. The uncorrected BIS spectrum for Ni.

absorbed or volatile materials. Additionally, BIS is not directly amenable to the analysis of insulating materials.

References

(1) Briggs, D., ed.; Handbook of X-Ray and Ultraviolet Photoelectron Spectroscopy; Heyden Press: London, 1972.

(2) Huang, J. T.; Rabalais, J. W., Electron Spectroscopy: Theory, Technique and Applications, Brundle, C. R.; Baker, A. D., eds.; Academic Press, London; 1978; 198.

(3) Kunz, C., Photoemission and the Electronic Properties of Surfaces; Feuerbacher, B., ed.; J. Wiley and Sons: New York, 1978; 501.

(4) Lindau, I.; Spicer, W. E., Synchrotron Radiation Research; Winick, H.; Doniach, S., eds.; Plenum Press: New York, 1980; 159.

ANGULAR DEPENDENT X-RAY PHOTOELECTRON SPECTROSCOPY

Use

In conventional XPS analysis, the angle at which photoemitted electrons are collected by the electron energy analyzer is fixed relative to the surface plane. Angular dependent XPS (ADXPS) refers to the systematic variation of the angle of emission, or photoelectron "take-off" angle in order to enhance surface sensitivity. By decreasing the take-off angle, the surface to volume signal ratio increases in a predictable manner. Surface enhancements of up to 10 times are possible at very shallow angles.

Sample

One can perform ADXPS on any solid material which has a relatively flat surface and is at least 2 mm in diameter.

Principle

The principle of angular dependent XPS is shown diagramatically in Figure 1. Photoemitted electrons traveling through a solid have a relatively high probability of interacting with neighboring atoms and undergoing inelastic energy loss processes before leaving the sample surface. Those electrons that are able to escape without a loss in energy appear in the XPS spectrum at discrete energies above a smooth background of inelastically scattered electrons. The distance that an electron can traverse through the solid before experiencing inelastic scattering is defined as the mean free path (λ). The value of λ is dependent upon the electron kinetic energy and varies from 0.5–5 nm over the energy range of 100–2000 eV but is independent of the trajectory of the electron for most samples. However, the sampling depth which is measured normal to the plane of the sample surface is highly dependent on the electron trajectory. As can be seen from Figure 1, the sampling depth (d) is at a maximum for a corresponding take-off angle (Θ) of 90°. If λ is known, d can be calculated for any take-off angle by the following expression:

$$d = \lambda \sin \Theta$$

Thus, an enhancement in surface sensitivity by a factor of greater than 10 can be obtained by changing the take-off angle from 90° to 5°.

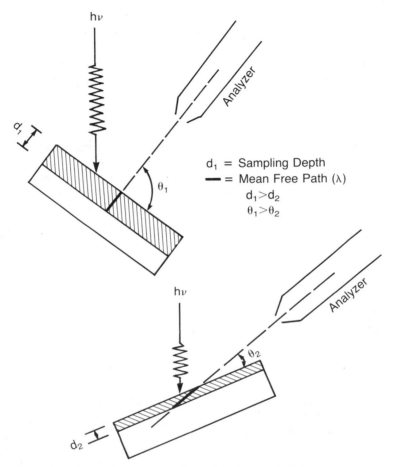

Figure 1. Schematic representation of the basic ADXPS experiment.

Applications

The technique of angular dependent XPS has found its greatest utility in the studies of ultra-thin films or overlayers, and in the analysis of monolayer and submonolayer quantities of adsorbed materials on a solid substrate. These types of studies are particularly relevant in the fields of catalysis, semiconductors, coatings and corrosion. Inherent in this technique is the ability to perform nondestructive depth profiling and interface analysis. For example, the capability has allowed the unambiguous elucidation of the Si/SiO_2 interface which is of critical importance in the understanding of semiconductor device performance. Additionally, the bonding geometries of atoms or molecules adsorbed on single crystal surfaces have been extensively studied via the photoelectron diffraction effect (diffraction of photoelectrons from crystal planes) for a wide variety of systems including epitaxial growth of monolayers on GaAs.

Intensive research efforts have focused on methods for the production of ultra-thin (<1.5 nm) dielectric layers for use in the fabrication of microelectronic devices through MIS (metal-insulator-semiconductor) technology. Silicon nitride is ideal for this application due to its favorable dielectric properties and imperviousness to diffusion of impurities. Figure 2 repro-

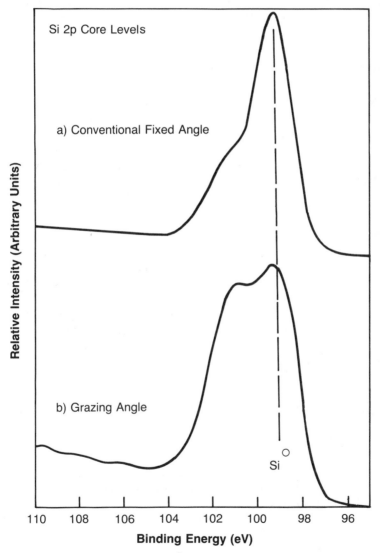

Figure 2. XPS spectra of a thin (<15 Å) Si_3N_4 film on a silicon substrate recorded (a) using a conventional fixed angle and (b) with a grazing take-off angle.

duces the XPS spectra of the Si2p region of a thin silicon nitride film formed on a silicon substrate by an ion beam method. The spectrum taken at the conventional fixed take-off angle indicated detection of the silicon substrate at $E_b = 99$ eV and Si_3N_4 as evidenced by the shoulder to the high binding energy side of the substrate peak. By employing a shallow take-off angle, it was possible to enhance the signal originating from the Si_3N_4. Detailed analyses at various take-off angles revealed a film thickness of 1 nm and the presence of non-stoichiometric nitrides of silicon at the uppermost layer of the film. This type of information is invaluable in order to correlate device performance to film quality.

Limitations

In addition to those limitations which exist for angle integrated XPS analysis several additional factors must be considered for angular dependent XPS. Because the angle of electron emission must be precisely defined, the technique requires that the sample be relatively flat. Moreover, angular dependent XPS requires a longer data acquisition time relative to a comparable angle integrated analysis. Thus, potential sample damage as a result of radiation exposure must be closely monitored.

References

(1) Fadley, C. S., Electron Spectroscopy: Theory, Techniques and Applications, Brundle, C. R.; Baker, A. D., eds.; Academic Press: New York, 1978.
(2) Plummer, E. W.; Eberhardt, W., Advances in Chemical Physics; Vol. 49; Prigogine, I.; Rice, S. A., eds.; J. Wiley and Sons: New York, 1982; 533.
(3) Riggs, W. M.; Parker, M. J., Methods of Surface Analysis; A. W. Czanderna, ed.; Elsevier Publishing: New York, 1975; 103.

SCANNING TUNNELING MICROSCOPY

Use

Scanning tunneling microscopy (STM) can provide high resolution images of surface morphology on the atomic scale. In addition, it can be used to study surface electronic structure.

Sample

Samples to be examined by STM must be in the solid phase, relatively flat and conductive. These can include cleavage faces of crystalline material, deposited thin films of semiconductors or metals and biological material mounted on a suitable substrate.

Principle

In many respects the principle of operation in STM is analogous to the operation of a profilometer. Both techniques use a fine tip to probe the degree of surface roughness. In the case of STM this "surface roughness" corresponds to contours (or the corrugations) in the sample's surface on the atomic level; the radius of the 'fine tip' or needle can have atomic dimensions. The major difference between the two techniques is in the phenomenon which gives rise to the ability of STM to measure surface contour. In order to attain such resolution, STM takes advantage of a quantum mechanical effect that is a direct consequence of the wave-like behavior exhibited by particles in motion. For example, consider a particle of mass m and total energy E approaching a potential barrier of height V such that $E < V$. In the classical case the particle will turn back or be reflected when it encounters the potential barrier. That is to say, classical mechanics predicts that it is impossible for the particle to penetrate the barrier. Quantum mechanics on the other hand predicts that there is a finite probability that a particle (such as an electron) will penetrate or tunnel through the barrier. The probability of the particle tunneling through the barrier (referred to as the transmission probability) falls off exponentially with distance from the barrier, i.e., $\exp(-K/r)$ where K is a constant. Likewise, if one considers a solid's surface there is a finite probability that a bound electron can tunnel through the surface barrier and hence be found beyond the surface boundary. In the language of quantum mechanics one would state the following: if an electron is described by a wave-function $\phi(x,t)$ then the probability density, $D(x,t)$, of finding the associated electron in the region between x and $x + dx$ is

$$D(x,t)dx = \phi^*(x,t)\phi(x,t)dx$$

If dx is the region beyond the surface potential barrier then one can see that there is a finite probability density (and a finite probability flux) that the electron will be found in this space. Experimentally this is verified by what appears to be a very simple experiment; i.e., if a potential difference is maintained between two electrodes (characterized by work functions ϕ_1 and ϕ_2) that are separated by a short distance current can flow between the two electrodes. This is a direct consequence of the tunnel effect and results from an overlap of the electronic wavefunctions associated with the two separate electrodes. The current density, j, associated with electron tunneling through this barrier for small voltages V can be expressed as:

$$j\alpha\ V\exp(-rA\sqrt{\phi})$$

where A is a constant on the order of several eV, ϕ the barrier height and r is the gap distance.

Now consider Figure 1 which reproduces schematically a profilometer (Figure 1a) and a STM (Figure 1b). The mechanical device depicted in Figure 1a employs a diamond tip stylus with a radius of ~ 5 μm which rests on the sample surface. As the stylus traverses the sample the tip experiences an upward and/or downward displacement. This displacement is sensed by a differential type transducer and a profile or roughness curve is generated. For STM the tip radius of the stylus is on the order of atomic dimensions. The stylus is supported by three mutually orthogonal piezoelectric elements. Elements x and y scan the tip over the surface which initially is brought to within several Angstroms of the sample surface; the z-piezoelectric element acts as a control—it both measures and maintains a constant tunneling current while the tip is scanned over the sample surface. This is accomplished by the application of a feedback voltage to the z-transducer. The stylus and the sample comprise the two electrodes and the gap or tunneling barrier can be vacuum, gas, or liquid. In order to maintain a constant tunneling current as the probe tip follows the contours of the surface, it follows from the expression for j that the z-position of the tip relative to the surface (r) must be varied. In this manner, a contour map of the surface is produced. The vertical displacements experienced by the tip are analyzed by computer aided image processing techniques. The image of the surface contour is usually displayed as a gray scale image.

Referring again to the expression for j, it is clear that the tunneling current is a very sensitive function of the distance r between the probe tip and the surface; i.e., it varies exponentially. Changes in the tunneling current can vary one order of magnitude with a change in r of only 1 Å. Hence, STM is very sensitive to the contour exhibited by a surface and the vertical resolution is on the order of 0.01 Å. The lateral resolution is dependent on the tip radius and can be on the order of 2 to 5 Å or roughly on the order of atomic distances in the surface plane.

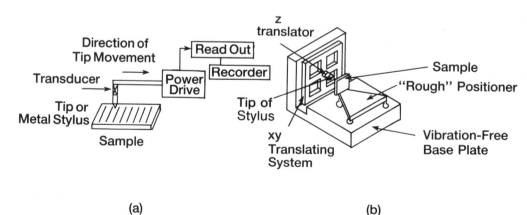

(a)   (b)

Figure 1. Schematic representations of: (a) a mechanical profilometer; (b) scanning tunneling unit.

Thus far the use of STM has been discussed as a means of imaging surface topography by scanning the probe tip across a surface while maintaining a constant tunneling current. However, the magnitude of the tunneling current does not only depend inversely on the separation between sample and probe tip, but also on the electronic structure. After all, as stated earlier, it is the overlap between wave functions associated with the two "electrodes" that is responsible for generating the tunnel current. By measuring current vs. voltage for nearly constant tip position STM becomes a surface sensitive spectroscopic technique that can be used to generate images of surface electronic states. By selecting the polarity of the tunneling voltage one can selectively probe the filled electronic states of the sample below the Fermi level (negative bias applied to the sample) or probe the empty states above the Fermi level (positive bias applied to the sample). This mode of STM operation provides information that is complementary to XPS/UPS and BIS respectively.

Applications

A major application of STM is the characterization of surface structure on an atomic scale.

One of the first investigations involved the "real space" determination of the 7×7 surface reconstruction of Si (111). This particular reconstruction has been the object of several theoretical and experimental efforts over the years. An adjusted gray scale tunneling image of a portion of that reconstruction is shown in Figure 2 (see Reference 1). The diamond shaped 7×7 unit cell is outlined. The areas of lighter intensity represent the top most atoms of the Si surface. Note the six fold symmetry of these intensity maxima about the unit cell corners (intensity minima).

Two other features are apparent in this micrograph. Arrow #1 shows a surface point defect (a missing atom). It's origin is still uncertain. Arrows #2 show a surface misfit dislocation. This defect becomes more obvious by noting the disregistry of atoms (intensity maxima and minima) on both sides of the defect itself.

STM has also proven useful in the study of biological materials (for example, the bacteriophage ϕ 29). For such studies, freshly cleaved pyrolytic graphite substrates were used.

Limitations

As of this writing, STM could be considered a technique still in the "infancy" stage. With this in mind, the degree of automation and the ease of data acquisition and presentation has not reached the advanced stage of development that is very common in most branches of microscopy. In addition, ongoing investigations are underway in order to determine to what

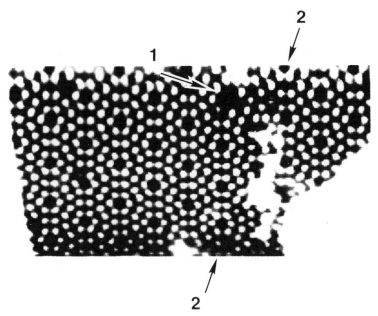

Figure 2. Missing atoms (#1) and surface misfit dislocations (#2) observed on the Si (111) 7 × 7 surface.

extent the surface is modified as a result of tip-surface interactions and to routinely produce sharp and stable probe tips.

References

(1) IBM Journal of Research and Development, 30(4), 1986.
(2) IBM Journal of Research and Development, 30(5), 1986.
(3) Simmons, J. G., J. Appl. Phys., 34, 1963, pp 1803–1973.
(4) Tromp, R. M.; Hamers, R. J.; Demuth, J. E., Science, 234, 1986, pp 304–309.

Acknowledgment

Figure 2 is reprinted from: Demuth, J. E.; Hamers, R. J.; Tromp, R. M.; Welland, M. E., IBM Journal of Research and Development, 30, 1986, 401; Copyright 1986; with permission of International Business Machines Corporation.

CHAPTER 9

Thermal Analysis

Edith A. Turi, Yash P. Khanna, Thomas J. Taylor

THERMOGRAVIMETRIC ANALYSIS (TG)

Use

Thermogravimetric analysis or thermogravimetry (TG) is used to determine changes in sample weight, which may result from chemical or physical transformations, as a function of temperature or time.

Sample

Materials in the solid or liquid state can be examined with a minimum sample size of about 1 mg.

Principle

TG is a technique which measures and automatically records changes in weight as a function of temperature. Isothermal TG measures the weight change as a function of time at a constant temperature. The data obtained provides information concerning the thermal stability, composition and decomposition behavior of the original sample.

The TG instrument, in conjunction with differential thermal analysis (DTA) and mass spectrometry, also provides a technique to investigate reactions such as dehydration, polymerization and decomposition.

TG and DTA are often used in conjunction, because they complement each other. It is possible to obtain DTA and TG results simultaneously on a single sample using combination thermal analytical instruments. Inert, vacuum, high pressure, or reactive environments, and various heating rates can be applied for the experiments ranging from $-150°C$ to $2400°C$. Sample size normally used is between 20 and 1000 mg, but the weight can be increased to about 12 g depending on the density.

A schematic of a thermogravimetric analyzer is given in Figure 1.

Applications

Areas of application include purity determination, screening of additives (e.g. plasticizer, filler, flame retardants etc.), determination of thermal and oxidative stability, evaluation of moisture, volatiles and residues, determination of catalyst performance, composition of blends and copolymers, flammability characteristics and reaction kinetics. Magnetic transitions can also be evaluated. Four examples are presented which indicate the versatility of this technique:

1. An example of the dehydration and decomposition of calcium oxalate monohydrate is given in Figure 2. A three-step weight loss pattern is observed as the temperature is raised at $10°C/min$: the loss of hydrate water, the loss of CO to form $CaCO_3$, and finally the formation of CaO residue by elimination of CO_2. The accompanying derivative or rate of weight loss curve shows the temperatures related to the most rapid rate of loss (or gain) for each weight change, and is valuable in correlating TG and DTA data.
2. Figure 3 illustrates the decomposition of an ethylene vinylacetate copolymer in an inert atmosphere. The initial weight loss, with the most rapid

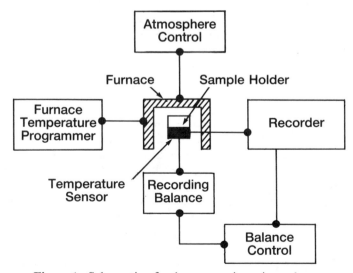

Figure 1. Schematic of a thermogravimetric analyzer.

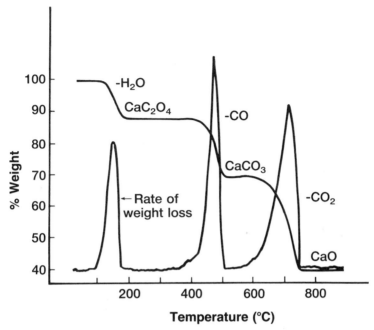

Figure 2. Calcium oxalate monohydrate ($CaC_2O_4 \cdot H_2O$). TG in nitrogen (10°C/min. heating rate).

loss occurring at ~380°C, corresponds to the degradation of the vinyl-acetate segment of the copolymer while the second step shows the depolymerization of the ethylene portion.

3. The electrical conductivity of polyacetylene doped with acceptors e.g. BF_4^-, PF_6^- and $CF_3SO_3^-$ decreases upon exposure to above ambient temperatures as a result of decomposition. In order to predict the extent of decomposition and thus conductivity of the doped polymer under a variety of time-temperature (t-T) conditions, isothermal TG experiments have been carried out at various temperatures (Figure 4). The results fit very well into a kinetic model for diffusion controlled processes, and the following expression has been obtained for the BF_4^-/polyacetylene system: $\ln(-\ln\alpha) = 14.92 - 6314/T + 0.45 \ln t$ where α is the weight fraction remaining at temperature T (K) and time t (hrs). Therefore, initial attempts can be made to predict the extent of decomposition under various t-T conditions.

4. When a magnet is placed around a ferromagnetic sample in the TG unit, an apparent gain in weight is observed. Upon reaching the Curie temperature (i.e. transition of ferro to paramagnetic) the apparent weight gained initially is lost. Figure 5 reveals the Curie temperature of the amorphous METGLAS® $Fe_{78}B_{17}Si_5$ at about 400°C. Upon continued

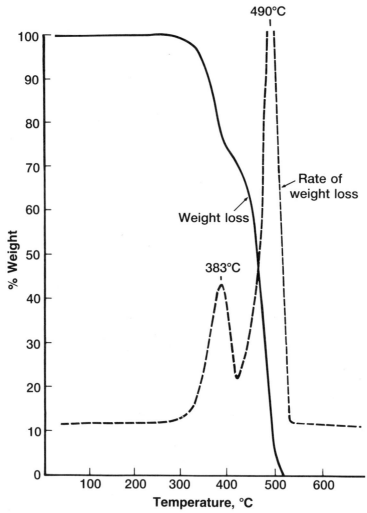

Figure 3. Ethylene vinylacetate copolymer. TG in argon (10°C/min. heating rate).

heating, multiple magnetic transitions are shown as the devitrification of the amorphous alloy takes place.

Limitations

This technique gives information on the overall weight change for a material, but does not provide information as to the nature of products evolved. Thus, an understanding of the decomposition process becomes difficult.

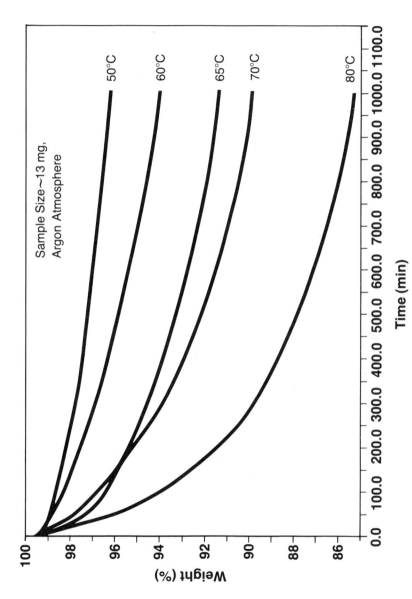

Figure 4. Isothermal TG of BF_4^- doped polyacetylene.

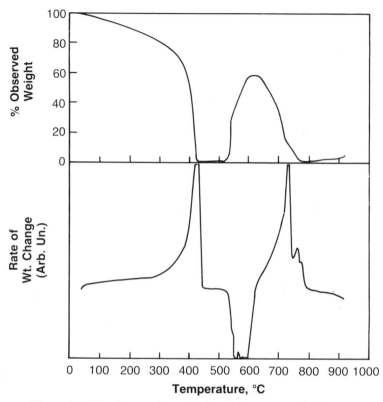

Figure 5. TG of amorphous $Fe_{78}B_{17}Si_5$ in a magnetic field.

References

(1) Keattch, C. J.; Dollimore, D., An Introduction to Thermogravimetry; Heyden: New York, 1975.

(2) Wendlandt, W. W., Thermal Methods of Analysis; Third Ed., Vol. 19, Wiley: New York, 1985.

(3) Wendlandt, W. W.; Gallagher, P. K., Instrumentation, Chapter 1 in Thermal Characterization of Polymeric Materials; Turi, E. A. ed., Academic Press: New York, 1981; pp 1–90.

Acknowledgments

Figure 1 is reprinted from: Wendlandt, W. W., Thermoanalytical Techniques; in Handbook of Commercial Scientific Instruments; Vol. 2, Marcel Dekker Inc.: New York, 1974; 144; by courtesy of Marcel Dekker Inc.

Figure 4 is reprinted from: Khanna, Y. P.; Taylor, T. J., Polymer Engineering and Science, 27, 1987, 766; with permission of Society of Plastics Engineers.

DIFFERENTIAL THERMAL ANALYSIS (DTA) AND DIFFERENTIAL SCANNING CALORIMETRY (DSC)

Use

Differential thermal analysis can be used to detect the physical and chemical changes which are accompanied by a gain or loss of heat in a material as its temperature is increased, decreased or held isothermally. Differential scanning calorimetry can provide quantitative information about these heat changes.

Sample

Samples may be examined in the solid or liquid state. Some information may be obtained with samples as little as 0.1 mg, but quantitative studies usually require at least 1 mg.

Principle

DTA and DSC are techniques for studying the thermal behavior of materials as they undergo physical and chemical changes during heat treatment. When a substance is heated, various chemical and physical transformations occur involving the absorption of heat (endothermic process) or evolution of heat (exothermic process). DTA measures the temperature difference arising between a sample and an inert reference material as both are heated at a constant rate in the same environment, thereby indicating endotherms and exotherms, and the temperature ranges over which they occur. The DSC technique, on the other hand, measures the amount of heat that is involved as a material undergoes either an endothermic or exothermic transition.

Various environments (vacuum, inert atmosphere or controlled gas composition) and heating rates (from 0.1°C/min to 320°C/min) can be employed for temperatures ranging from -150°C to 2400°C. Simultaneous DTA and TG can be obtained on a single sample.

Qualitative and quantitative measurements are determined with speed and efficiency through automated data analysis and reporting systems.

A schematic of a differential thermal analysis system is given in Figure 1.

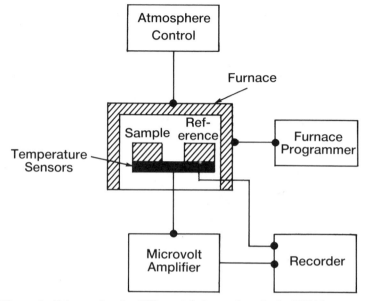

Figure 1. Schematic of a differential thermal analysis (DTA) system.

Applications

The DTA and DSC techniques can be used to investigate the thermal properties of a variety of materials, and are particularly useful tools in the characterization of organics, polymers, biological materials, inorganics and amorphous alloys. Some applications are: qualitative and quantitative evaluation of phase transformations such as glass transition, melting, crystallization; study of polymerization, decomposition and curing processes including a kinetic description; determination of thermal and processing histories, simulation of processing conditions and crystal growth.

Figures 2 and 3 reveal the polymerization isotherms of pivalolactone monomer using temperature (T) and initiator concentration (C) variables, respectively. The polymerization peak is analyzed for cumulative area (i.e. % conversion) as a function of time (t). Data analysis provides the following relationship: $\ln t_{0.75} = 3534/T - 4.157(C) - 6.801$, where $t_{0.75}$ is the time in minutes, required for 75% polymerization. Thus, the polymerization rate of pivalolactone monomer, e.g. in composite fabrication, may be optimized by the temperature and/or initiator concentration variables.

Figure 4 shows a DSC scan of an amorphous metal alloy. The first small endotherm (290°C) represents the Curie temperature (magnetic transition), and is followed by a heat capacity change at 490°C due to the glass transition. Devitrification (crystallization) occurs at 535 and 605°C.

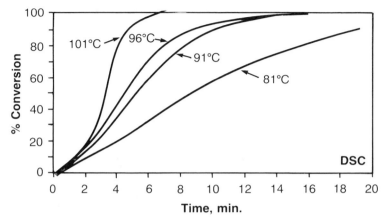

Figure 2. Polymerization isotherms for pivalolactone containing 0.15% initiator.

High temperature DTA readily reveals factors which affect the crystallization of a gallium gadolinium garnet crystal. Figure 5 indicates a noncon-
gruent, two phase crystallization transition obtained on cooling after the
sample has been overheated ~60°C above the melting point. However, an
overheating of ~15°C above the melting point produces a single phase crystal structure.

Figure 6 shows the effect of purity on DSC melting peak shapes for benzoic acid. The presence of 2.8% impurity depresses the melting point, significantly broadens the melting range and decreases the heat of fusion. A
computerized program for the DSC determines the purity automatically.

The kinetics of polymerization of diacetylene can be developed by determining the transition peak temperature of the exothermic reaction directly

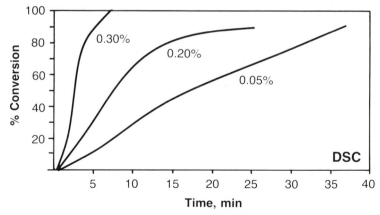

Figure 3. Polymerization isotherms for pivalolactone at 91°C: initiator concentration dependence.

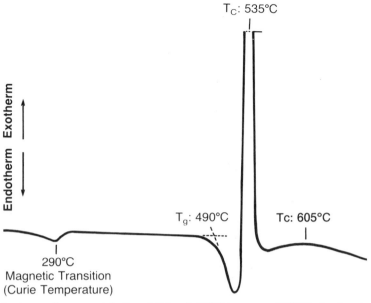

Figure 4. Amorphous metal alloy (ribbon). DSC in argon (20°C/min. heating rate).

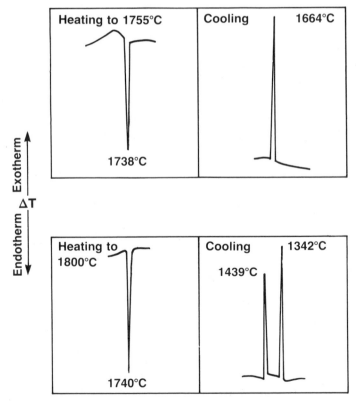

Figure 5. Gallium gadolinium garnet. High temperature DTA in purified He (25°C/min. heating rate).

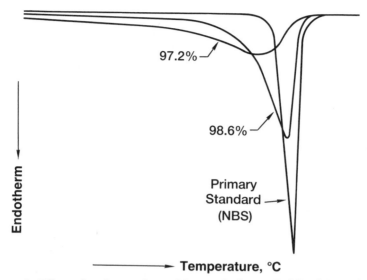

Figure 6. Effect of purity on the melting peak shapes (DSC) of Benzoic Acid.

in the instrument as a function of heating rate. Figure 7 illustrates the differences in the peak shape and temperature for three rates.

Figure 8 is a DSC scan of poly(ethylene terephthalate) (PET) indicating the glass transition (T_g), "cold" crystallization (T_c) and melting (T_m) (left to

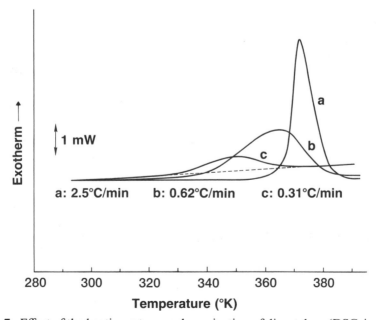

Figure 7. Effect of the heating rate on polymerization of diacetylene (DSC, in argon).

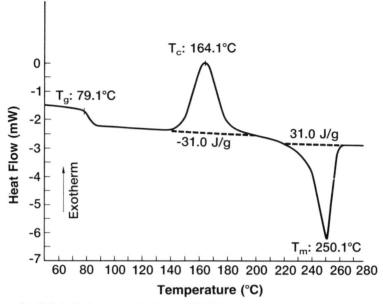

Figure 8. Poly(ethylene terephthalate) DSC in Argon (20°C/min. heating rate).

right). The automatic data analysis system has also printed out the heats of crystallization (ΔH_c) and melting (ΔH_f).

Limitations

Analysis of solutions becomes difficult using DTA/DSC techniques. Also if the change in thermal energy per unit time is small, as for example in secondary relaxations, cross-linked or semi-crystalline polymers, gelation processes, etc., the applicability of these techniques is sometimes limited.

References

(1) Hemminger, W.; Hohne, G., Calorimetric Fundamentals and Practice; Verlag Chemie: Basel, 1984.

(2) MacKenzie, R. C. ed., Differential Thermal Analysis; Vol 1&2, Academic Press: London, 1970, 1972.

(3) Pope, M. I.; Judd, M. D., Differential Thermal Analysis; Heyden: New York, 1977.

(4) Wunderlich, B., The Basis of Thermal Analysis; Chapter 2 in Thermal Characterization of Polymeric Materials, Turi, E. A. ed., Academic Press: New York, 1981; pp 92–228.

Acknowledgments

Figure 1 is reprinted from: Wendlandt, W. W., Thermoanalytical Techniques; in Handbook of Commerical Scientific Instruments; Vol. 2, Marcel Dekker Inc., New York, 1974; 4; by courtesy of Marcel Dekker Inc.

Figure 6 is reprinted from: DSC-2 Manual, Perkin-Elmer; Feb. 1976, p 4–5; with permission of the Perkin-Elmer Corp.

THERMOMECHANICAL ANALYSIS (TMA) AND DILATOMETRY

Use

TMA and dilatometry are the techniques used for measuring the dimensional changes in materials as a function of temperature and time.

Sample

A variety of samples can be analyzed e.g. plaques, fibers, films, powders, metallic ribbons etc.

Principle

The TMA technique employs a weighted probe resting on the sample which measures the linear displacement of the probe as the material is program heated or cooled. The movement of the sample is translated into an electrical signal by a linear variable differential transformer, and recorded as a function of temperature, or of time (isothermal measurement). Interchangeable probes are available to measure penetration, expansion, elongation or shrinkage over a temperature region of $-150°C$ to $1500°C$. Predetermined force can also be applied to study dimensional changes under load.

Dilatometry provides dimensional changes over a wide temperature region (RT to $2000°C$) and in various environments (e.g. inert, air or vacuum).

A schematic of a thermomechanical analyzer is given in Figure 1.

Applications

These techniques are quite versatile. Some applications are: evaluation of dimensional stability of polymers (fibers, films etc), metal alloys, ceramics, crystals, composites, etc., information about mechanical anisotropy, orien-

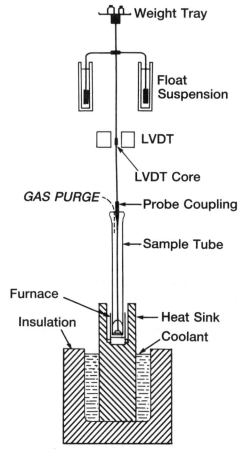

Figure 1. Schematic of a thermomechanical analyzer.

tation and morphology, detection of transitions, and evaluation of thermo-set materials. The study of sintering phenomena in materials is also an important area.

In Figure 2 an expansion mode curve is given for an epoxy molding com-pound. A T_g of 150°C was determined for the material by extrapolation. Also determined was a low temperature expansion coefficient (α_1) in the temper-ature range of 60–110°C, and a high temperature expansion coefficient (α_2) in the temperature range of 180–230°C.

A sintering study of a commercial zirconia/yttria ceramic by dilatometry is shown in Figure 3. The sintering process, reflected by shrinkage, occurs in the 1000–1500°C region in two distinct steps. This information combined with electron microscopy provides details on the sintering mechanism. Other processing variables that lead to maximum densification can also be identified.

Figure 4 shows the suppression of shrinkage with increasing load for a commercial poly (ethylene terephthalate) (PET) fiber.

Limitations

Experimental problems restrict the applicability usually to solids. In addi-tion, the samples can undergo creep simultaneously with their normal dimensional changes.

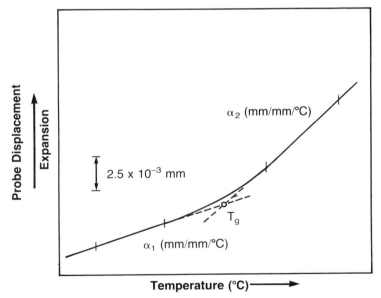

Figure 2. Linear expansion coefficient of an epoxy thermoset. TMA in He (5°C/min. heating rate).

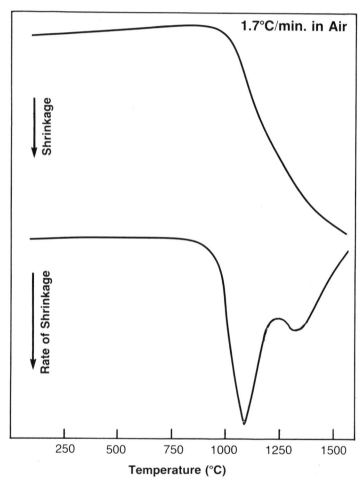

Figure 3. Dilatometry of a commercial zirconia/yttria ceramic.

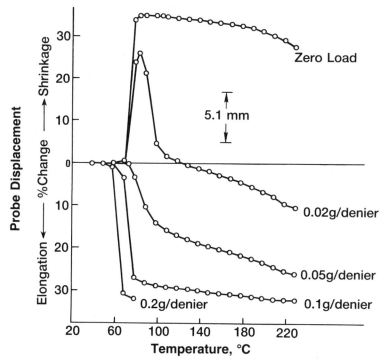

Figure 4. Effect of load on dimensional changes of commercial PET fibers [TMA in He (10°C/min. heating rate)].

References

(1) Jaffe, M., Fibers; Chapter 7 in Thermal Characterization of Polymeric Materials; Turi, E. A. ed., Academic Press: New York, 1981; pp 709–785.

(2) Maurer, J. J., Elastomers; Chapter 6 in Thermal Characterization of Polymeric Materials, Turi, E. A. ed., Academic Press: New York, 1981; pp 572–704.

(3) Prime, R. B., Thermosets; Chapter 5 in Thermal Characterization of Polymeric Materials: Turi, E. A. ed., Academic Press: New York 1981; pp 435–563.

Acknowledgment

Figure 1 is reprinted from: TMS-1 Manual, Perkin-Elmer; Feb. 1971, pp 2–2; with permission of the Perkin-Elmer Corp.

THERMAL CONDUCTIVITY

Use

Thermal conductivity data as a function of temperature is useful in measuring the thermal transport properties of materials.

Sample

DSC based methods are available which use small samples e.g., 0.25″ diameter and thickness of $\geqslant 0.05″$. However, for most other methods samples must normally be in the form of a disc with dimensions of at least approximately 2″ diameter and thickness ½″.

Principle

Most commercial units are of the comparative type, with the unknown specimen being sandwiched between two identical reference materials. A temperature difference is established between the top and bottom of this "stack" and, at equilibrium, the temperature difference between the top and bottom of the sample is measured. This enables the thermal conductivity to be calculated using a heat flow equation which determines the change in conductivity of the two reference materials when a sample is present. A number of absolute units rely on a similar geometry except that heat flow meters are used in place of standards to measure the heat transfer through the sample.

Applications

Thermal conductivity can be determined for a wide range of materials including polymers, ceramics, alloys and composites. One example is presented to indicate the use of this technique. Figure 1 shows the thermal conductivity of METGLAS® alloy 2605S-2, measured across the ribbon width.

Limitations

The measurements are lengthy, accuracy of the data is normally ± 15%, liquids in general cannot be analyzed as a function of temperature, and for polymers the variation of thermal conductivity with temperature is not well understood.

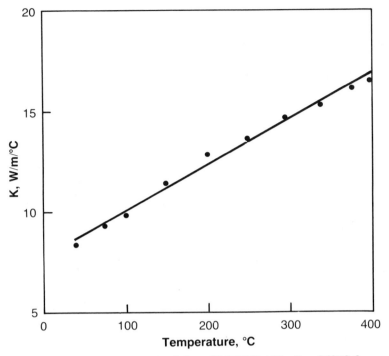

Figure 1. Thermal conductivity of METGLAS® alloy 2605S-2.

References

(1) Chiu, J.; Fair, P. G., Thermochim, Acta., 34, 1979, 267.
(2) Tye, R. P., Thermal Conductivity; Vol. I & II, Academic Press: New York, 1969.
(3) Wendlandt, W. W.; Gallagher, P. K., Instrumentation; Chapter 1 in Thermal Characterization of Polymeric Materials; Turi, E. A., ed., Academic Press: New York, 1981; pp 1–90.

DYNAMIC MECHANICAL ANALYSIS (DMA) AND SONIC MODULUS

Use

DMA provides information on changes in the viscoelastic properties of materials as a function of temperature, time or frequency at a constant oscillatory strain. Sonic modulus test measures the dynamic elastic modulus at room temperature and at a constant frequency.

Sample

Commercially available instruments allow a variety of sample shapes to be analyzed i.e. fibers, films and sheets (1–200 mils thick). Also a wide range of materials can be analyzed e.g. polymers, composites and metal foils.

Principles

In DMA an oscillatory strain is applied to the sample in the bending or tensile mode of deformation as a function of temperature or time; the frequency (0.01–200 Hz) and strain (0.03–1.7%) are preselected and maintained constant throughout the analysis. The resultant oscillatory stress which lags the applied strain by a phase angle δ for a visco-elastic material, is monitored by a transducer. Three parameters are calculated, dynamic storage (elastic) modulus E', dynamic loss (viscous) modulus E'', and the dissipation or damping factor, $\tan \delta = E''/E'$. These properties can be measured from $-150°C$ to $500°C$. A schematic of a dynamic mechanical analyzer is given in Figure 1.

In the sonic modulus test, a sonic pulse at 5 kHz is transmitted along the material at RT and the flight time between transmitting and receiving transducers is measured. The sonic velocity thus obtained, provides the dynamic modulus of elasticity.

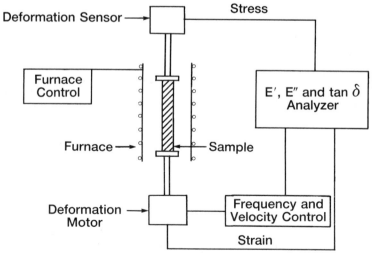

Figure 1. Schematic of dynamic mechanical analysis (DMA) instrumentation.

Applications

Some common applications of DMA include phase homogeneity in polymeric blends, alloys and copolymers, curing of thermosets, stiffness and anisotropy in composites, and selection of proper materials for specialized applications e.g. acoustics, automotive and aerospace. The role of the polymer matrix with respect to the high temperature stiffness of graphite fiber composites is readily obtained by DMA (Figure 2). This information helps in the selection of a proper matrix for a particular application. In Figure 3, the usefulness of DMA is shown in studying the molecular mixing between a polyester carbonate (COPEC) and a poly (ethylene terephthalate) (PET); a ratio of 67/33 of COPEC/PET appears to be slightly phase segregated as suggested by a trace of PET transition at 85°C, whereas the 90/10 COPEC/PET material appears to be homogeneous, exhibiting only one transition at 156°C.

Sonic modulus provides information on the molecular orientation in fibers, paper and films. For example, the increase in sonic modulus with increasing load in the case of polyethylene fibers, has been attributed to the amorphous phase orientation occuring under load. Interestingly, no such effect is observed for KEVLAR® polyamide fibers as illustrated in Figure 4.

Limitations

It may not be possible to examine the samples in their original morphology e.g. powders, since these have to be molded. Commercially available

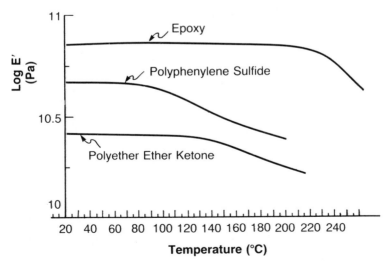

Figure 2. DMA of graphite fiber composites.

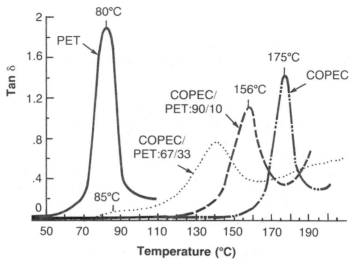

Figure 3. Polymeric alloys of COPEC/PET.

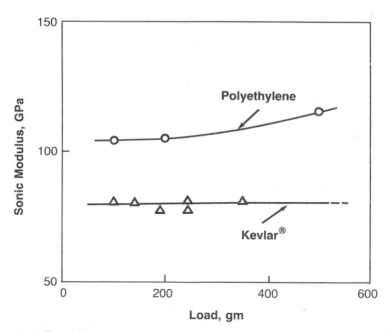

Figure 4. Effect of load on the sonic modulus of polyethylene and Kevlar® fibers.

dynamic mechanical spectrometers are limited to an upper frequency of 200 Hz. Sonic modulus measurements are normally restricted to room temperature.

References

(1) Ferry, J. D., Viscoelastic Properties of Polymers; Third Ed., Wiley: New York, 1980.

(2) Marayama, T., Dynamic Mechanical Analysis of Polymeric Material; Elsevier: New York, 1978.

(3) Read, B. E.; Dean, G. D., The Determination of Dynamic Properties of Polymers and Composites, Wiley: New York, 1978.

DIELECTRIC THERMAL ANALYSIS (DETA)

Use

This technique measures the dielectric properties of a material as a function of temperature and frequency. Specifically, the dielectric constant (ε'), dielectic loss (ε'') and the dissipation factor (tan δ) are measured over the $-150°C$ to $+300°C$ region at a frequency of 20 Hz to 100 kHz.

Sample

A thin film or sheet is required with a surface area of 5 cm^2.

Principle

In DETA, an oscillating electrical field with a preselected frequency, is applied to the sample as a function of temperature. For an insulating material such as plastic or rubber, the electrical polarization lags behind the applied field by a phase angle δ. The electrical displacement analysis using bridge techniques provides the dielectric constant (ε'), dielectric loss (ε'') and the dissipation factor (Tan $\delta = \varepsilon''/\varepsilon'$).

Applications

Parameters such as ε', ε'' and tan δ as a function of frequency and temperature are important in selecting proper materials for electrical applications. The technique is highly sensitive for detecting dipolar or ionic species in a material. Figure 1 shows the dielectric data for a rubber material coated with a small amount of hydroxy derivative; the relaxation peak at 0°C is

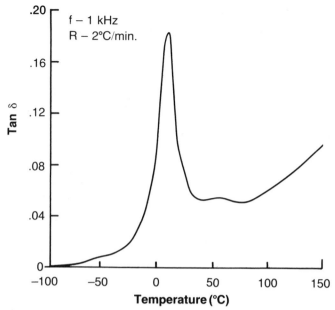

Figure 1. Dielectric dissipation factor vs. temperature for a modified SBR rubber.

found to be missing in the absence of hydroxy derivative. In addition to the practical aspects, DETA in conjunction with DMA allows for a detailed analysis of the molecular motions.

Limitations

Samples may not be analyzed in their original shape e.g. powders, since a molded piece is needed for the measurements. Also liquid samples are sometimes difficult to handle.

References

(1) Hedig, P., Dielectric Spectroscopy of Polymers; Wiley: New York, 1971.
(2) McCrum, N. G.; Read, B. E.; Williams, G., Anelastic and Dielectric Effects in Polymeric Solids; Wiley: New York, 1967.

Acknowledgment

Figure 1 is reprinted from: Polymer Laboratories Brochure; with permission of Polymer Laboratories Ltd.

CHAPTER 10

The Viscosity and Molecular Weight of Polymers

Abraham M. Kotliar, Milton E. McDonnell, Eugene K. Walsh

RHEOLOGY OF FLUIDS

Use

Rheometers generally employ simple flow geometries, the most common being capillary flow (Figure 1), cone-and-plate (Figure 2), and Couette (Figure 3), to measure the viscoelastic characteristics of liquids, solutions, and melts. A recent addition to the above line of rheometers is a squeezing flow apparatus which has the capability of measuring biaxial elongational viscosity (Figure 4).

Sample

Sample size depends on the instrumentation used and the scope of the problem; 100 g is usually required, but measurements can sometimes be made with as little as 10 g. Measurements can be made from 20 to 380°C with capillary flow and -140 to 380°C with the cone-and-plate geometry.

Principle

The simplest type of rheological behavior is a Newtonian fluid described by the relationship

$$\tau = \eta\dot{\gamma}$$

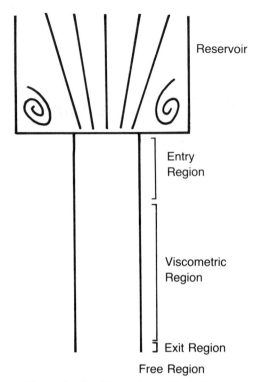

Figure 1. Capillary viscometer geometry.

where τ is the shear stress in Pascals (Pa), η the coefficient of viscosity in Pascal seconds (Pa·s) and $\dot{\gamma}$ the shear rate in reciprocal seconds. The viscosity of a Newtonian fluid is independent of shear rate and strain history. Many real fluids show both a pseudoplastic and elastic behavior (see Figure

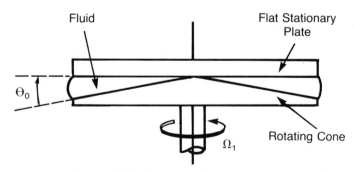

Figure 2. Basic cone-and-plate geometry.

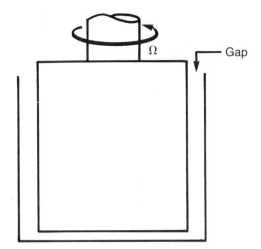

Figure 3. Rotational viscometer, Couette type.

5). These fluids often can be characterized by an empirical power law type of relationship

$$\tau = k\dot{\gamma}^n$$

where k and n are constants for a particular fluid. The parameter k is a measure of the consistency of the fluid: the higher k, the more viscous the fluid. The parameter n is a measure of the degree of non-Newtonian behavior: the greater its departure from unity, the more pronounced the non-Newtonian characteristics of the fluid.

There are a number of other phenomenological, molecular, and empirical descriptions of the behavior of fluids which are applicable for many rheological problems related to elasticity, stress overshoot, and molecular entan-

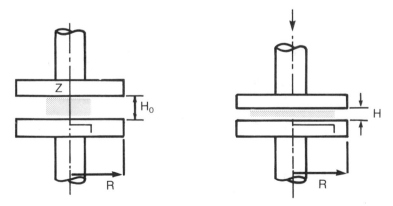

Figure 4. Lubricated squeezing flow.

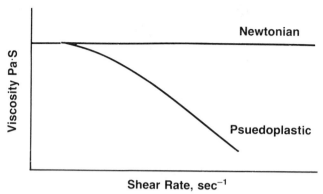

Figure 5. Typical melt flow curve.

glements. However these are beyond the scope of this brief introduction. For most engineering applications the Newtonian and power law descriptions are adequate.

Applications

Rheological measurements are used to determine processing conditions, design dies for extrusion and spinning, and for process control. As a research tool, viscosity measurements have been used to determine thermal stability of polymers, cross-linking reactions, interchange rates, hydrolysis and polymerization rates, melt heterogeneities and the characteristics of the molecular weight distribution in flow situations.

Limitations

Most polymer processing involves both shear and elongational flows. Because of the melt elasticity of polymer melts, these two flow processes are simply related as is the case for Newtonian fluids. Measurements of melt elasticity and elongational viscosity are extremely difficult to carry out.

References

(1) Bird, R. B.; Armstrong, R. C.; Hassager, O., Dynamics of Polymer Liquids; Second Edition, John Wiley: New York, 1987.
(2) Han, C. D., Rheology in Polymer Processing; Academic Press: New York, 1976.

(3) Petrie, C. J. C., Elongational Flows; Pitman: San Francisco, 1979.
(4) Tadmor, Z.; Gogos, C. G., Principles of Polymer Processing; John Wiley: New York, 1979.

MECHANICAL SPECTROMETRY

Use

A mechanical spectrometer measures the rheological and mechanical properties of a broad range of materials using either steady state or dynamic strains. The equipment has cone-and-plate, parallel plate and Couette geometries for melts and fluids. With solids such as fibers, films, rods or bars, one can employ torsion, tension-compression and three point bending. A number of deformation histories can be applied to the material in a temperature range of -140 to 380°C. Typical apparatus can provide four decades of strain frequency from 10^{-2} to 10^2 radians per second, with controlled strain and temperature sweeps.

Sample

The sample size depends on the type of tests required. Extensive mechanical property information can be obtained on films and fibers requiring less than one gram of material. Twenty five grams is generally sufficient for tests requiring bars and disks.

Principle

The mechanical spectrometer is a rigid apparatus which translates a selected deformation history to the motion of a servocontrolled motor through a microprocessor and a fixed geometry. The resulting forces are measured by two accurate normal and torque transducers. The measurements provide information on the steady state and dynamic elastic and loss modulus together with the phase angle and the complex viscosity. Sinusoidal, ramp and step strains can be programmed over four decades of rates.

Applications

Determination of the mechanical properties of solids and fluids over a broad range of temperatures and deformation rates provides detailed infor-

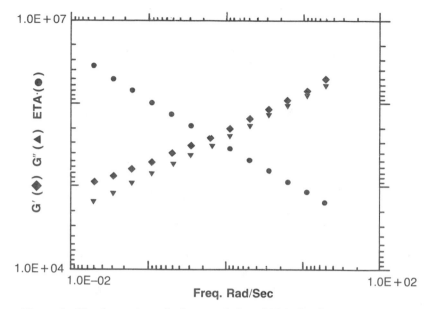

Figure 1. The dynamic melt characteristics of high density polyethylene.

mation on the various transitions the material undergoes as a function of time and temperature. This information is essential in many end-use applications of plastics and fibers. With polymer melts and other fluids, knowledge of the viscosity and elasticity over a wide range of shear rates provides unique material characterization about its molecular weight and molecular weight distributions and how they may affect processing and end use applications. Figure 1 shows typical results obtained on a high density polyethylene blow molding resin. The high storage modulus, G′, and pseudoplastic nature of the complex viscosity provides information on the melt elasticity and the high molecular weight fraction of the molecular weight distribution.

Limitations

The measurements generally require extended time periods. This becomes problematic at elevated temperatures where the material can undergo significant changes during the measurement.

References

Ferry, J. D., Viscoelastic Properties of Polymers; Third Edition, John Wiley: New York, 1980.

MOLECULAR WEIGHT OF POLYMERS

Molecular weight is one of the most fundamental properties of any molecule. Almost all physical properties of polymers systematically change as the molecular weight is altered. Unlike pure substances of small molecules, synthetic polymer samples have a range of molecular weights. For this reason, there is no such quantity as "the molecular weight"; there are certain average molecular weights or molecular weight distributions. There follow three different methods that determine different averages of molecular weight and a discussion on crosslink density, which describes the molecular weight between crosslinks. Elsewhere in this book are three other subsections of interest for finding molecular weight distributions: gel permeation chromatography, which separates molecules by size; photon correlation spectroscopy, which can infer the distribution of molecular weight from the correlation function; and field flow fractionation, which separates components in a manner dependant on the type of field applied.

COLLIGATIVE PROPERTIES

Use

The number average molecular weight, which is the mass of solute divided by the number of molecules, can be measured by colligative properties of dilute solutions. The determination is absolute in the sense that it can be traced to basic physical measurements, as opposed to a relative measurement which compares some physical property of a molecule to the same physical property of a series of calibration standards. Absolute methods are not affected by branching or solvent-solute interactions. When the number average molecular weight is compared to another molecular weight average, the polydispersity of the sample can be assessed. The second virial coefficient, which characterizes the solvent-solute interaction, can also be determined.

Sample

All forms of materials in quantities as low as 100 mg can be evaluated. Small sample quantities, however, will require prior knowledge of solvent miscibility.

Principle

Colligative properties such as freezing point dispersion or boiling point elevation are directly related to the change in chemical potential of the solvent in a solution from its value in a pure state. Only two colligative properties can be measured with great enough sensitivity to be applicable to the relatively low molarity of polymer solutions—the osmotic pressure and vapor pressure of the solvent. The instruments used to make each measurement are the membrane osmometer and vapor pressure osmometer, respectively.

When a solution is separated from its pure solvent by a membrane permeable to the solvent, but not the solute, an excess pressure builds up in the solution to counter the reduced chemical potential of the solute molecules in the solution. When the osmotic pressure per concentration of dissolved material is extrapolated to infinite dilution, the value is the reciprocal of the molecular weight of the polymer. Since most polymeric materials contain some low molecular weight species which are smaller than the pores in the membrane, the resulting molecular weight average includes only species above the cutoff for the membrane. These methods provide the most accurate methods of determining number average molecular weight, provided that the material can be dissolved in a solvent which does not dissolve the membrane.

In a vapor pressure osmometer, the reduced vapor pressure of the solvent in a solution decreases the rate of evaporation and results in a higher temperature for the solution than for the pure solvent. This temperature change is on the order of a few thousandths of a degree, so it must be measured with great precision with an electric bridge circuit. This technique does not extend to molecules of as high a molecular weight as measurable by membrane osmometry, but it has fewer restrictions on usable solvents.

Applications

1. The number average molecular weight of a polymer can be measured by colligative properties. For example, a function of the osmotic pressure of polyisobutylene is plotted as a function of concentration in two different solvents in Figure 1. The intercept of either line is related to the molecular weight of the sample.
2. The second virial coefficient which characterizes the solvent-solute interaction can also be obtained. In the example of Figure 1, dissolving the polymer in a poor solvent, benzene, results in a negative slope while a good solvent, cyclohexane, gives a positive slope. The slope is directly proportional to the second virial coefficient.
3. Many physical properties such as solubility, flexural fatigue resistance, and impact strength are related to the polymer molecular weight. Optical and dielectric properties which are influenced by oligomers and unreacted monomers correlate more closely to number average molecular weight than the weight average; however, the entire molecular weight distribution must be known to fully describe the properties.

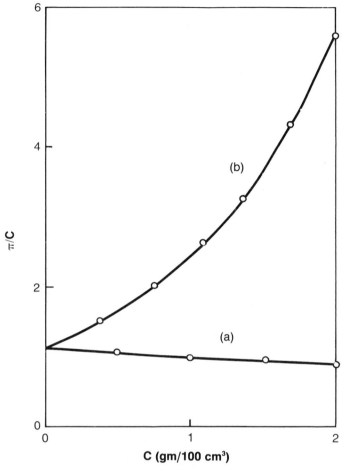

Figure 1. Osmotic pressure-concentration ratio vs. concentration for (a) benzene (b) cyclohexane solutions of polyisobutylene.

4. Impurity levels in the polymer can be assessed by osmometry because they produce the same colligative effects as polymers.

Limitations

All colligative methods become insensitive at sufficiently high molecular weight since the dilute solutions have an insufficient number of particles to produce a detectable change in the property being monitored. For membrane and vapor pressure osmometry, this is about 1,000,000 and 30,000 molecular weight respectively. For membrane osmometry, a membrane is required that is permeable to the solvent but impermeable to the solute. Generally, the pores are of such size that molecules less than 5,000 molecular weight pass through the membrane and are therefore removed from the average. Vapor pressure osmometry, in contrast, is sensitive to the monomer and oligomers as well as low molecular weight impurities.

References

(1) Morawetz, H., Macromolecules in Solution; John Wiley and Sons: New York, 1965; pp 161–196.

(2) Tanford, C., Physical Chemistry of Macromolecules; John Wiley and Sons: New York, 1961; pp 180–274.

Acknowledgment

Data in Figure 1 is reprinted from: Flory, P. J., J. Am. Chem. Soc. 65, 1943, 377; Copyright 1943; with permission of the American Chemical Society.

VISCOSITY OF POLYMER SOLUTIONS

Use

Viscosity determinations of dilute polymer solutions are frequently used to estimate molecular weight. Unlike classical light scattering and colligative property measurements, which determine the molecular weight directly, viscosity measurements compare the viscosity of a solution of the polymer with the viscosity of solutions of calibration standards. If the polymer has the same composition, component distribution, branching, and conformation as the standards, viscosity methods provide a rapid means of accurately assessing the molecular weight. Likewise, viscosity measurements are useful in arranging a series of polymers of uniform composition and branching structure in order of changing molecular weight. The molecular weight determined for a heterogeneous sample is called the viscosity average molecular weight. It is an average whose weighting depends on the strength of the solvent-solute interactions, not just the polymer molecules. The viscosity average molecular weight usually corresponds to the weight average of the sample more closely than the number average molecular weight.

Sample

All forms of materials, in quantities as low as 10 mg, can be evaluated. Small sample quantities, however, will require prior knowledge of solvent miscibility.

Principle

The viscosity of a fluid characterizes the internal friction of one layer of fluid flowing over another. The viscosity of a liquid can be found from measurements of the time it takes for a specified volume to flow through a vertical capillary of known length and diameter. When polymer chains are added to a solution, the macromolecules greatly retard the flow; that is, they increase the viscosity. Viscosity can be increased either by putting in a higher concentration of polymers of an equivalent molecular weight or the same concentration of polymers with a higher molecular weight. Molecular weight of some polymers has been correlated to the solution viscosity at a specified concentration in a given solvent at a certain temperature. Such correlations are widely used in quality control but are of very limited use for research.

Viscosity measurements are made more accurate by extrapolating results to low polymer concentration. The specific viscosity of a polymer is the ratio of difference between the viscosities of the solution and solvent to that of the solvent. If the specific viscosity divided by the concentration, the reduced viscosity, is extrapolated to infinite dilution, the resulting value is called the intrinsic viscosity. Despite the name, this quantity is not a viscosity. Note that it has the dimensions of reciprocal concentration, not viscosity. Einstein showed that intrinsic viscosity could be related to the density of the polymer molecule. Since the density of most polymer conformations decreases uniformly with molecular weight, the intrinsic viscosity can be used as an indicator of molecular weight. Such empirical relations, called Mark-Houwink equations, apply only to one polymer-solvent system at a specified temperature. If the appropriate Mark-Houwink equation is known, viscosity measurements become a rapid means of determining molecular weight.

Applications

1. Viscosity measurements are useful for assessing molecular weights of systems that have previously been calibrated. Figure 1 shows an example of such a calibration for nylon 6 samples where reduced viscosity in m-cresol is compared to independently determined number and weight average molecular weight. The molecular weight averages of future samples can be estimated from the reduced viscosity measurements.
2. Viscosity measurements are useful for showing qualitatively how molecular weight changes with aging, processing, or end use exposure even when no calibration has been made.
3. Estimates of the hydrodynamic volume of polymers dissolved in dilute solutions can be made from intrinsic viscosity measurements.
4. Many physical properties, such as melt viscosity, impact resistance and

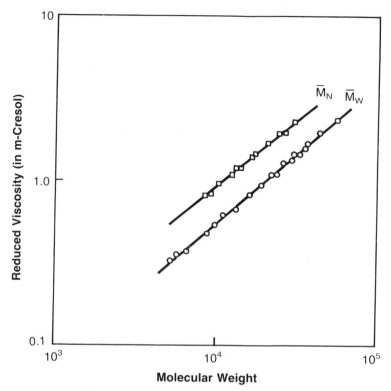

Figure 1. Reduced viscosity of nylon 6.

solubility, vary systematically with molecular weight. The viscosity average molecular weight is usually 10 to 20% below the weight average molecular weight.

Limitations

Since viscosity offers a relative method to determine molecular weight, it is necessary that the sample be similar in composition, conformation, component distribution, and branching structure to the calibration standards for the results to be accurate. Unfortunately, the measurements seldom indicate if any of these conditions are being violated.

References

(1) Carpenter, D. K.; Westerman, L., Polymer Molecular Weights, Part II; Slade, Jr., P. E., ed., Marcel Dekker: New York, 1975.

(2) Morawetz, H., Macromolecules in Solution; John Wiley & Sons: 1965; pp 299–314.

CLASSICAL LIGHT SCATTERING

Use

Light scattering can be used to determine the weight average molecular weight, that is, the average weighted by the mass concentrations of components, rather than the number concentration. In any heterogeneous sample, the weight average molecular weight is always larger than the number average, and the ratio of the weight to number average is called the polydispersity. Like the number average measured from colligative properties, light scattering provides an absolute, rather than relative, determination. By analyzing the amount of light scattered from solutions at different polymer concentrations and angles, the second virial coefficient and the radius of gyration of the molecule can also be determined.

Sample

All forms of materials in quantities as low as 10 mg can be evaluated. Small sample quantities, however, will require prior knowledge of solvent miscibility.

Principle

The oscillating electric field of a beam of light causes the electrons of molecules in its path to vibrate sympathetically and redirect some of the incident radiation. The amount of light scattered by polymer molecules is much greater than that from solvent molecules since the light scattered by the large number of electron oscillators of the polymer add together constructively. For particles that are small compared to the wavelength of light, the amount of scatter increases with the square of the molecular weight and is independent of the shape. Thus, light scattering measurements are more influenced by larger particles and give higher values than the number average. The high scattering from large particles also demands the complete removal of large contaminants from the sample since these would greatly increase the measured molecular weight. The scattering from particles larger than one twentieth the wavelength of light becomes more complex. Scattering measurements from these particles must first be extrapolated to zero angle before determining molecular weight and second virial coefficients. This can either be done in a photometer specially designed to make measurements only a few degrees from the transmitted beam, or by extrapolating measurements taken at a series of higher angles. The first method is much faster and gives a more accurate molecular weight determination. The alternate approach gives the radius of gyration of the polymer, a weighted average of its mass segments, in addition to the molecular weight and second virial coefficient. The schematic of a typical low angle laser light scattering photometer is given in Figure 1.

Applications

1. The weight average molecular weight of a polymer can be determined by light scattering. An example is shown in Figure 2 where a function of the light scattered at different angles (Θ) from solutions of different concentrations (c) is plotted. By extrapolating this line to both zero angle and zero concentration, the molecular weight of the polymer can be determined.
2. The second virial coefficient of the polymer, which characterizes the polymer-solvent interaction in the solution, is also obtained. This is related to the slope of the line connecting the measurements extrapolated to the low angle limit in Figure 2.
3. The radius of gyration of the polymer in the given solvent can be found. This comes from the slope of the line connecting the measurements extrapolated to the zero concentration limit in Figure 2.
4. Many physical properties such as hardness, softening temperature, and elongation at tensile break are related to the polymer molecular weight. Polymer melt viscosity and impact strength are more closely related to the weight average molecular weight than the number average although both properties are functions of the entire molecular weight distribution.
5. Particle sizes can be measured by light scattering.

Limitations

In order to obtain accurate molecular weight by light scattering, it is necessary to have samples that are clean, optically different from the solvent, and non-aggregating. Since many impurities scatter more light than polymer systems, it is necessary to have clean systems to assure that the molecular

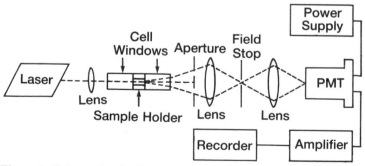

Figure 1. Schematic of a low angle laser light scattering photometer.

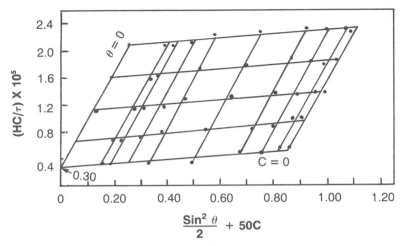

Figure 2. Light scattering from polymer solutions.

weight being determined is that of the polymer. If the refractive index increment is small, little or no light will be scattered. Finally, since light scattering averages are weighted toward the larger particles, the presence of concentration dependent aggregates causes difficulties.

References

(1) Huglin, M. B., ed., Light Scattering from Polymer Solutions; Academic Press: New York, 1972.
(2) Kerker, M., The Scattering of Light and Other Electromagnetic Radiation; Academic Press: New York, 1969.

CROSSLINK DENSITY

Use

Solvent swelling techniques define the crosslink density and/or molecular weight between crosslinks of cured materials. The method evaluates the effectiveness of various crosslinking agents. It also provides a measure of the optimum degree of cure needed for product end-use.

Sample

The samples must be solid. If the material is completely crosslinked, as little as one gram is sufficient for this measurement. Analysis time is dependent on how rapidly equilibrium swelling is attained.

Principle

The degree to which a crosslinked material swells in a given solvent is a function of its crosslink density and its mutual compatibility with the solvent. The higher the density of crosslink sites in a material the less its degree of swell when exposed to a solvent. A measure of the equilibrium volume of solvent absorbed by the crosslinked material at a given temperature and the degree of compatibility of the solvent with the material (the latter obtained by measuring the solvent-solute interaction parameter) provides the data needed to determine the crosslink density and/or the molecular weight between crosslinks in the material.

Precision

The precision of the method is primarily dependent on the accuracy of the measurement of the solvent-solute interaction parameter. This value is normally obtained by colligative property measurements on the material. Uncertainty in this measurement results in a precision of approximately 10 to 15% in molecular weight between crosslink calculations. Results reported in terms of swelling index usually vary by approximately 3%.

Applications

1. The molecular weight between crosslinks or the crosslink density of a gel can be determined. As an example, a series of polystyrene gels with dif-

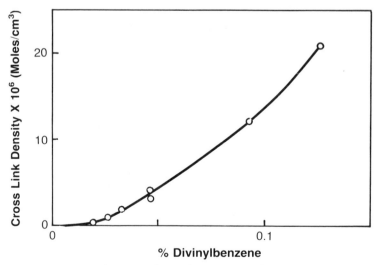

Figure 1. Cross link density of gels.

ferent amounts of divinylbenzene crosslinking agents were synthesized. Figure 1 shows the relation between the composition and the crosslink density of these samples.

2. The effectiveness of various crosslinking agents can be assessed.
3. The effect of processing conditions on gel formation can be measured.

Limitations

This method is restricted by the types of samples that can be studied and limited by the applicability of the theory. Samples must have sufficient structural integrity to be subjected to a variety of solvent conditions for measurements. Crosslink density of some samples, such as certain water soluble ionic gels, cannot be determined by this method since current theories do not correctly describe their swelling.

References

(1) Flory, P. J., Principles of Polymer Chemistry; Cornell University Press; Ithaca, New York, 1953; pp 347–398.
(2) Volkenstein, M. V., Configurational Statistics of Polymeric Chains; Timasheff, S. N. and Timasheff, M. S., trans., John Wiley & Sons: New York, 1963; pp 447–556.

Acknowledgment

Data in Figure 1 is reprinted from: Boyer, R. F.; Spencer, R. S., J. Polym. Sci., 3, 1948, 104; Copyright© 1948, with permission of John Wiley & Sons, Inc.

FIELD FLOW FRACTIONATION

Use

Field flow fractionation (FFF) is a series of separation techniques where a field is applied perpendicular to particles flowing through a channel so as to position particles with different characteristics in different flow lines. The extent of the particle field interaction determines the transport time through the channel and can produce good chromatographic resolution. A variety of external fields can be used—sedimentation, thermal, flow, and electrical— to produce separation based on particle size, thermal properties, or charge.

Generally, sedimentation field flow fractionation (SFFF) is most useful for characterizing particles; thermal field flow fractionation (TFFF) is most helpful in finding molecular weight distributions of polymers. Flow field flow fractionation (FFFF) and electric field flow fractionation (EFFF) are being developed to study distributions of particle size and charge, respectively. Since there is no separating matrix as in GPC, there is no possibility of unwanted chemical interactions with the media. At the same time, the GPC limitations of limited pore size and shearing of large macromolecules are avoided. When the field is systematically varied during the run, very wide ranges of particle sizes can be separated from a single sample. The physics of the interactions, and the geometry of the equipment are relatively straight forward; therefore, physical properties of eluting samples can, in principle, be directly calculated, rather than compared to calibration standards.

Sample

Typically, 50 μL of a 1% suspension are injected into a SFFF instrument. The flow is stopped while the particles reach their equilibrium position in the field. Then the flow is resumed and the particles elute in times that depend on their size and density difference from the solvent. In many cases, this time is less than an hour when the field is varied to speed up the differentiation of the particles. Injection volumes of macromolecular solutions tend to be larger and more dilute but the total amount of sample is comparable. Elution times for broad samples by FFF techniques, other than sedimentation, tend to be longer since methods for the systematic programming of the field are not readily available; the good resolution of early peaks results in a long retention of components in slow flow lines.

Principle

Separation takes place in a long narrow rectangular channel which is wide compared to the particles. As fluid flows through this channel, it adopts a parabolic flow profile with the highest speed in the center and lower speeds towards the walls. The external field partitions the particles so that ones with similar properties have a characteristic distance from the cell wall and therefore a specific flow time through the channel. This flow time is proportional to the ratio of the field-induced velocity of the particle to the translational diffusion coefficient. The channel wall prevents the particle from moving with the field-induced velocity as it would in a centrifuge. It is, however, this calculated quantity which determines the flow lines carrying the particle.

In SFFF, the channel is located in a ring, equidistant from the axis of the ultracentrifuge. Special connectors permit solvent to enter, flow through, and exit the spinning channel. The separation time is proportional to the product of the particle volume, its density difference from the solvent, and the acceleration of the centrifuge. Large particles go to one edge of the channel and traverse in slow flow lines. Particles which are heavy compared to the solvent go to the outer wall while those light compared to the solvent go to the inner wall.

The channel is sandwiched between a high and low temperature wall, producing a thermal gradient perpendicular to the direction of solvent flow in TFFF. The separation is proportional to the ratio of the translational to thermal diffusion coefficient times the thermal gradient across the cell. Pressurizing the system leads to larger temperature gradients while keeping the solvent in the liquid state.

With FFFF, the channel is made with sides that are permeable to the solvent but not the solute. A transverse flow field is created so the solvent moves both through and across the channel. Elution times of suspended particles are related to their size. Because of available membranes, the technique has been limited primarily to particles in aqueous solvents. An electric field is applied across the channel in EFFF. The retention is a function of the electrophoretic mobility of the particles.

Steric field flow fractionation results when the size of the particle approaches the width of the separation channel. Then all the particles are close to the channel wall, and it is the particle thickness which determines the retention time. The elution times of large and small particles are reversed between regular and steric FFF, so it is imperative that the whole separation be carried out in only one mode. Any type of separating field previously mentioned can be used in steric FFF. In Figure 1 is given a schematic of a field flow fractionation instrument where the separating field can be any of those described.

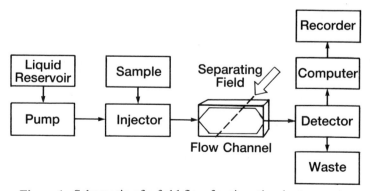

Figure 1. Schematic of a field flow fractionation instrument.

Applications

1. SFFF characterizes particle size distributions of colloidal systems. Figure 2 shows the comparison of elution times for two latex samples. It is apparent that each sample has two major components; however, in the good sample, the two components are closer in size and the larger particles of the good material are more uniform than in the bad sample.
2. With information on operating conditions and particle shape and density, elution time values can be converted to a particle size distribution. Because of the high resolution, particles with as little as 10% difference in size can be separated.
3. Systematic variation of the field lets SFFF give in less than an hour the particle size distribution of an emulsion or dispersion where sizes cover several orders of magnitude.
4. Molecular weight distributions can be obtained by SFFF for ultra high molecular weight polymers such as polyacrylamide. Such polymers degrade in a GPC column.
5. Molecular weight distributions of polymers can be obtained by TFFF. This is particularly useful for polymers that strongly interact with GPC media or are soluble in solvents that attack GPC columns such as sulfuric acid.
6. Particle size distribution of emulsions where each phase has similar density, can be obtained with FFFF.
7. Differently charged macromolecules can be separated by EFFF. The technique is more useful for biological systems where monodisperse protein

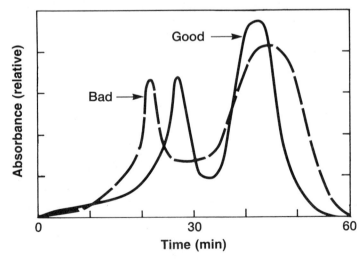

Figure 2. Sedimentation field flow fractionation data.

components are encountered than for synthetic materials which have a distribution in both molecular weight and charge.

Limitations

While FFF is a powerful set of techniques, experimental conditions are complex and optimal procedures have not been determined. Separation is ultimately limited by the strength of field that can be applied to the channel. Because solvent is pumped through an ultracentrifuge rotor, spinning over an electric motor in SFFF, safety considerations limit the operating temperature and appropriate solvents. Size distribution information is based on the assumption that the particles are spherical and have uniform density. Since the separation is based on both size and density, no separation results if the particles have insufficient density difference from the solvent. For this reason, SFFF is not applicable to low or medium molecular weight polymers.

Good polymer fractionation has been demonstrated by TFFF. Since there is no good theory to relate thermal and translational diffusion coefficients, this equipment does not permit direct calculations of molecular weight from elution time as is done with a sedimentation or flow field. Thus, calibration standards like those used in GPC are needed. FFFF does not have as good resolution as SFFF and is further restricted by available membranes. For this reason, it has not been as widely applied. EFFF has been developed even less fully.

References

(1) Caldwell, K. D., Modern Methods of Particle Size Analysis; Barth, H. G., ed., John Wiley & Sons: New York, 1984; pp 211–250.

(2) Giddings, J. C.; Graff, K. A.; Caldwell, K. D.; Myers, M. N., Polymer Characterization; Crover, C. D. ed., American Chemical Society: Washington, DC, 1983; pp 257–269.

CHAPTER 11

Physical Properties of Particles and Polymers

Milton E. McDonnell, Eugene K. Walsh

GENERAL METHODS FOR PARTICLE SIZE ANALYSIS

Use

General methods for particle size and size distribution analysis covered in this section include light scattering, the Coulter principle, optical gradient techniques and air permeability methods. Elsewhere in this book are subsections describing photon correlation spectroscopy, field flow fractionation, and gel permeation chromatography which provide similar information for many systems. The sizes covered by these techniques range from 1 nm to 250 μm. Cement, ceramic, glass dye, paint, emulsion, and pharmaceutical product performance are some of the systems affected by particle size considerations.

Sample

The amount of sample required for evaluation depends to some extent on the breadth of distribution of particle sizes in the sample. In general, light scattering photometry, Coulter Counter and optical gradient techniques require 10 to 100 milligrams of sample. The Sub-Sieve method requires an amount of 1 cm^3 of the actual material, excluding voids.

Principle

Light scattering principles are discussed in the section on classical light scattering. Since the scattered intensity can be modeled in terms of particle volume as well as molecular weight, the procedures to measure either one are similar. The general description of the light scattered and absorbed by large spherical particles in a dielectric is mathematically described by Mie's theory but experimental measurements are difficult. The situation is simplified as particles get smaller and the index of refraction of the continuous medium is close to that of the scatterers.

The Coulter principle determines a number-volume distribution of particles suspended in an electrically conductive liquid. The suspension flows through a small aperture having immersed electrodes on either side. Each passing particle displaces electrolyte within the aperture producing a voltage pulse proportional to the particle volume. These pulses are amplified, scaled, and counted using pulse height analysis. A schematic of a Coulter type instrument is given in Figure 1.

Optical gradient techniques employ a photometric measure of sedimenting particles. In a homogeneous suspension all particles will settle at a uniform rate, and the time of settling for a particle of a specific size can be calculated from Stokes' law. A measure of the decrease in turbidity with time, caused by slower sedimentation of particles of successively decreasing size, reflects the distribution of these particles. This is accomplished by an equation which considers the relationship between the light absorbed, particle size, and number.

The Sub-Sieve Sizer employs an air permeability method for measuring the average size of powdered solids. A current of air maintained at a given pressure is forced through a bed of powder packed to known degrees. A calibrated flowmeter is used to measure the air flow at various compaction lev-

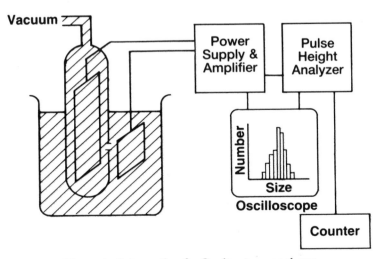

Figure 1. Schematic of a Coulter type analyzer.

els. The readout of the flow meter is converted by means of a calculator chart to the average particle diameter in the powdered sample.

Precision

Light scattering photometry techniques, when applicable, measure particle sizes in the 20 to 300 nm range with a precision estimated to be ∼10%. The Coulter Counter has given results equivalent to those obtained by other techniques when the materials being evaluated are chemically inert. Materials capable of being affected by the electrolyte medium required in this method have shown size deviations of 20% or more. Optical gradient techniques have been compared relatively infrequently with other size measuring methods. In specific instances of comparisons, principally with pigments, size values within ±5% have been obtained after corrections for differences in the type of average size applied. The Sub-Sieve Sizer technique has resulted in good average size correlations with other techniques in some applications and poor correlations in others. In general, microscopical methods are the basic, most direct means for evaluating small particle sizes and, as such, remain the prime methods by which the precision of all others will be compared. Some of the microscopical techniques are described in the Image Analysis article in this book.

Applications

Size and size distributions of particles are important in such diverse areas as:

 1. Catalyst performance
 2. Ceramic and alloy properties
 3. Mineral processing
 4. Pesticide and herbicide activity
 5. Contaminants in high purity liquids
 6. Emulsion stability
 7. Bubbles formed in liquids
 8. Paint performance
 9. Drug effectiveness
10. Atmospheric pollutant dispersal

An example is assessing the effect of the particle size distribution of a trace, 10–100 μm, contaminant in a colloidal suspension. Figure 2 shows the particle size distribution measured for one sample with a Coulter Counter. The sub-micron colloidal particles are too small to be observed by

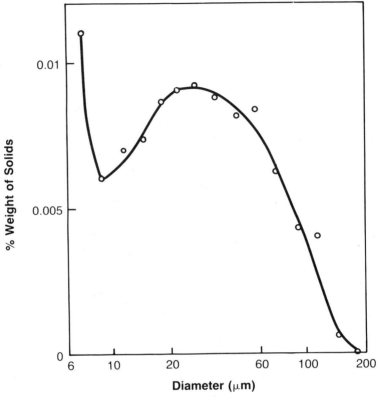

Figure 2. Size distribution of contaminant particles in a colloidal suspension from Coulter type analyzer data.

the detector. Performances of samples with different contaminant compositions were measured and correlated to the contaminant distribution.

Limitations

If particles must be suspended in a fluid dissimilar to that of their application environment when particle size measurements are made, the state of aggregation may change. Rough handling of suspensions may also damage fragile particles or aggregates. Each technique of particle size characterization has its own weaknesses that ideally should be inconsequential for the system being sized. The Coulter principle, for example, demands the particles be suspended in an electrolyte solution but does give a direct measure of particle volume. Other techniques usually interpret results in terms of particle size but spherical shape is assumed in the analysis. The differentiating of the effects of particle shape and distribution of particle size is difficult in any technique not classifying discrete particles.

References

(1) Allen, T., Particle Size Measurement; Chapman and Hall: London, 1981.
(2) Barth, H. G., ed., Modern Methods of Particle Size Analysis; John Wiley and Sons: New York, 1984.
(3) Barth, H. G.; Sun, S.-T., Anal. Chem., 57, 1985, 151R–175R.

PHOTON CORRELATION SPECTROSCOPY

Use

Photon correlation spectroscopy can measure the average size of particles or macromolecules in a liquid that is transparent to the incident light. In many cases, measurements can be made in the native state of the system without applying any external disturbing force. A distribution of particle sizes can also be extracted. The size that is measured is an effective hydrodynamic radius which is the size of a solid sphere that would diffuse at the same rate as the particle in question. For many polymers this size can be converted to a molecular weight.

Sample

All forms of materials in quantities as low as 10 mg can be evaluated. Small sample quantities, however, will require prior knowledge of solvent miscibility.

Principle

In classical light scattering, a solution appears to scatter a constant amount of light at a given angle, but this is not actually the case. The intensity fluctuates about the average value as the molecules continuously move about. At times one scatterer will be positioned relative to another to give constructive interference at the detector. As these particles move relative to one another, the interference will eventually become destructive and the scattered intensity will decrease. The rate of change of the scattered intensity thus becomes a measure of the rate of relative motion of scatterers in the solution.

Photon correlation spectroscopy characterizes these fluctuations with a correlation function that is computed by a dedicated computer. This correlation function is then related to an appropriate model of molecular or colloidal dynamics. The experimental setup is diagrammed in Figure 1.

The most prominent fluctuations usually result from the diffusion of the scatterers. In this case, the correlation function measures the diffusion coefficient of the species. Einstein showed that for spheres at sufficient dilution, the diffusion coefficient at a given temperature in a fluid of specified viscosity measures the particle radius. Mathematical techniques separate the correlation function into contributions from different components of the population. In some systems measurements are needed at different scattering angles to differentiate between fluctuations caused by translational diffusion and other dynamic processes.

Applications

1. Particle size and size distributions of emulsions, latexes, and sols can be measured by photon correlation spectroscopy. As an example, Figure 2 shows the distribution of sizes in an alumina sol that has been aged for one month. The small peak corresponds to the narrow distribution of primary particles, while the broad peak shows aggregates of the primary particle.
2. The kinetics of aggregation can be followed. This is demonstrated in the alumina sols previously discussed. When the sol was first made, all the particles had a radius close to 4 μm. On aging, aggregating particles formed a second component to the distribution which, with time, increased in both mean size and distribution width. Kinetic parameters can be calculated from these changes.

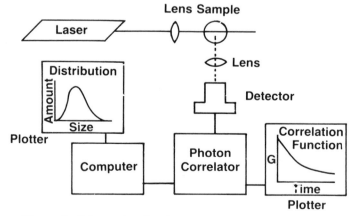

Figure 1. Schematic of a photon correlation spectrometer.

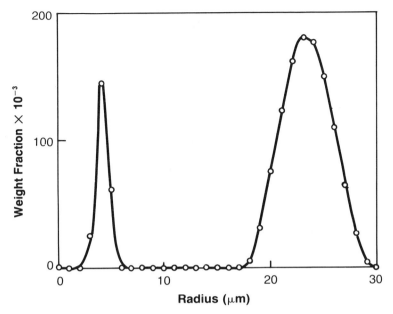

Figure 2. Particle size distribution of an alumina sol determined by photon correlation spectroscopy.

3. The molecular weight distribution of a polymer sample can be estimated if the relation between polymer size and molecular weight is known.
4. Shape information about larger asymmetric colloids can be obtained since correlations will also result from their rotational as well as translational motion.

Limitations

Since photon correlation spectroscopy measures the fluctuation of concentration about its equilibrium value, it is necessary that the particles remain suspended in the fluid and undergo negligible sedimentation during the course of the measurement. The particle size that violates this criteria decreases with increasing difference between the density of the particle and its suspending solution. As with all light scattering techniques, it is necessary to have a difference in the refractive index between the particle and its medium.

References

(1) Chu, B. Laser Light Scattering; Academic Press: New York, 1974.
(2) Dahneke, B. E., Measurements of Suspended Particles by Quasi-Elastic Light Scattering; John Wiley and Sons: New York, 1983.

(3) Jamieson, A. M.; McDonnell, M. E., Probing Polymer Structures; Koenig, J. L., ed., American Chemical Society: Washington, DC, 1979; pp 163–206.

(4) Pecora, R., ed., Dynamic Light Scattering; Plenum Press: New York, 1985.

GAS ADSORPTION

Use

Gas adsorption techniques define surface area, pore size and pore size distributions of solid materials. These parameters are used to obtain information on the catalytic activity, adsorptivity, permeability, filterability and granular compressibility of solids. Such measurements are of value in characterizing the products in the cosmetic, ceramic, pigment, petroleum, abrasive, drug and fertilizer fields.

Sample

The sample must be a solid or capable of being solidified at the temperature of liquid nitrogen. Sample quantity depends on the magnitude of its surface area and its density. Surface areas as low as 0.1 m^2/g have been successfully defined on sample quantities as low as 200 mg.

Total analysis time is dependent on the rapidity at which equilibrium is attained. When the method employed is that of constant gas flow, equilibration is relatively rapid.

Principle

The amount of gas needed to form a monomolecular layer on the solid surface can be determined from measurements of the volume of gas adsorbed as the pressure is increased by small increments at constant temperature. The BET (Brunauer, Emmett, and Teller) equation relates the adsorbed gas volume, the applied pressure P and saturation vapor pressure P_f in the function form shown in Figure 1. From the slope and intercept of this plot, the volume of the gas forming a complete monolayer can be calculated. This in turn can be converted to the surface area of the solid by dividing by the thickness of the gas layer.

The rate at which the gas is then desorbed from a completely covered surface as the pressure is decreased maps out the distribution of pore sizes on the surface. This is because adsorbed gas is bound more tenaciously in pores with smaller radii. Adsorption theory relates the amount of the desorbed gas from a pore of a given size to the pressure change.

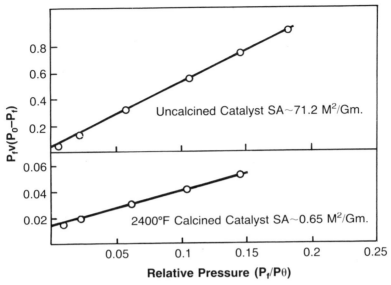

Figure 1. Adsorption of nitrogen at 77°K on catalyst before and after calcination.

A schematic of a continuous flow gas adsorption apparatus is given in Figure 2.

Precision

Surface areas of 1 m²/g or greater can be measured with an accuracy of ±2% or better. Measurement precision decreases with decreasing surface areas. Successful evaluations have, however, been accomplished on surface area standards in the 0.01 m²/g range with an accuracy of ±5%.

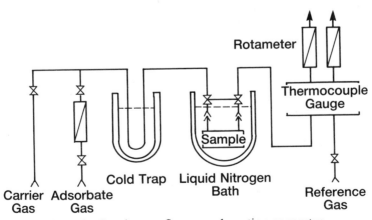

Figure 2. Continuous flow gas adsorption apparatus.

Applications

1. Surface area is an important factor in determining the activity of catalysts and adsorbants. As an example, Figure 1 shows a hundred fold decrease in surface area when a certain catalyst was heated to 2400°F.
2. Pore distribution information often gives insight into surface area measurements. Figure 3 gives an example. Gas desorption information for the catalyst discussed previously shows that the sharp decrease in surface area upon heating is the result of destroying a large number of pores.
3. The number of chemisorption sites on a catalyst can be measured since the gas adsorbed to these sites cannot be removed.
4. Estimation of mean particle size of sub-micron sized powders can be obtained from the surface area per mass and density. In contrast to microscopy, this method gives a mean directly without worries of forming representative samples of particles and is applicable to aggregated particles.
5. Decrease in surface area can be used to follow the removal of particle fines from a sample.

Limitations

Samples must be small enough to put in the measuring apparatus; that is, have a cross sectional area of a few square millimeters. At the same time,

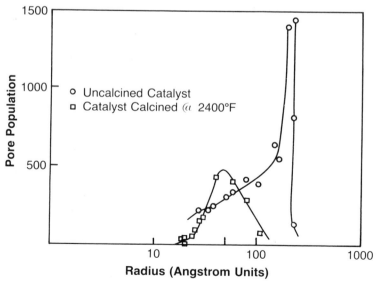

Figure 3. Pore distribution of catalyst before and after calcination from nitrogen desorption data.

they must offer enough surface area to give reliable results. The BET model of gas adsorption is assumed in the interpretation of the data to give surface areas. Since this model makes some nonphysical assumptions, such as the equivalence and independence of each adsorption site on the surface, the absolute numbers must be viewed cautiously.

References

(1) Adamson, A. W., Physical Chemistry of Surfaces; John Wiley and Sons: New York, 1982; pp 492–600.

(2) Allen, T., Particle Size Measurements; Chapman and Hall: London, 1981; pp 465–563.

(3) Gregg, S. O.; Sing, K. S. W., Adsorption, Surface Area and Porosity; Academic Press: New York, 1982.

(4) Lowell, S.; Shields, J. E., Powder Surface Area and Porosity; Chapman and Hall: New York, 1984.

MERCURY INTRUSION POROSIMETRY

Use

Mercury intrusion porosimetry defines the average pore size, pore size distribution, bulk density and skeletal density of non-compressible solids. The parameters obtained are used in the same capacity of those defined by gas adsorption techniques. The latter technique however is applicable to the 3 to 40 nm pore size range and may be applied to both soft and rigid solids. Mercury intrusion techniques measure pore sizes in the nm to 300 μm range. Both soft and rigid solids can be measured at and above $\sim$10 μm. Below this pore size this technique is applicable only to rigid particles. The parameters measured are important mainly to the fields of ceramics and metallurgy, although polymer foams have also been investigated by this technique.

Sample

The sample must be solid and non-compressible if pore measures of less than $\sim$10 μm are desired. Sample quantities required are nominally 0.3 to 1.0 grams. The latter range is dependent on the porosity of the sample being evaluated.

Principle

Pore structure analysis by mercury intrusion is based on measuring the volume of mercury forced into the pores of the sample as a function of pressure. The pressure at which intrusion into the pores occurs is inversely proportional to the pore diameter. Given appropriate values for the mercury surface tension and interfacial contact angle, the pore diameter at any applied pressure can be determined. A schematic of a mercury intrusion porosimeter is given in Figure 1.

Precision

The precision depends somewhat on the pore volume of the sample with larger pore volumes being somewhat easier to measure. In general, the measurements are accurate to within ~10%.

Applications

1. Material strength is related to pore structure. As an example, Figure 2 shows the pore distribution in a block of polystyrene foam. Different polystyrene foams are found to differ widely in the complexity of their pore structure.
2. The porosity of resins is related to their interaction with plasticizers.
3. Mercury porosimetry can be used to measure surface area and thus can

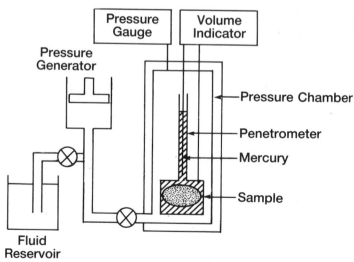

Figure 1. Schematic of a mercury porosimeter instrument.

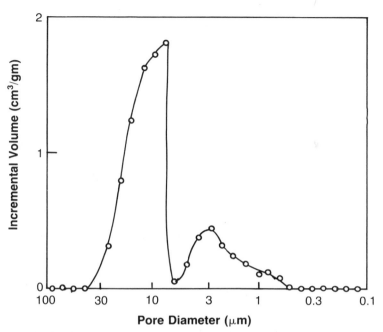

Figure 2. Distribution of pores in a polystyrene foam determined by mercury intrusion porosimetry.

be applied to similar problems mentioned in the subsection of gas adsorption.

Limitations

Mercury intrusion porosimetry cannot be applied to large, compressible, or fragile materials. Sometimes the differentiation between pores and interparticle spacing is not distinct. The technique cannot be used to investigate pores below 3 nm. The interpretation of the data in terms of pore size usually assumes that (1) all pores are cylindrical in shape and (2) the constant contact angle for mercury is independent of the sample. In many materials, hysteresis is produced from ink-bottle shaped pores or networks of interconnected channels.

References

(1) Allen, T., Particle Size Measurements; Chapman and Hall: London, 1981; pp 564–582.
(2) Gregg, S. J.; Sing, K. S. W., Adsorption, Surface Area, and Porosity; Academic Press: New York, 1982; pp 173–190.

(3) Lowell, S.; Shields, J. E., Powder Surface Area and Porosity; Chapman and Hall: New York, 1984, pp 205–216.

COHESIVE ENERGY DENSITY

Use

Cohesive energy density (CED) measurements are used to define solubility characteristics, compatibility, thermodynamic parameters, steric purity and regularity. Studies of this type are primarily used to select a solvent prior to all solution separations and molecular property evaluations. In general, it is employed in all problems concerned with compatibility of materials such as polymers with solvents, plasticizers and other polymers.

Sample

Any form of material can be evaluated in quantities as low as 50 mg. Quantity of material as well as analysis time will depend on the extent of work desired.

Principle

The cohesive energy density or the solubility parameter (square root of CED) is a measure of the energy associated with attractive forces between molecules (the molar internal energy per cubic centimeter). When the CED of two nonelectrolytes are equal, the probability of ideal mixing (complete miscibility) of materials is large. The matching of CED of various materials provides a preliminary guide to their mutual compatibility.

The most direct and accurate way to access the CED of a solvent is by use of its latent heat of vaporization (ΔH_v). When this parameter is not available in the literature, it can be estimated by use of the Clausius-Clapeyron equation. The energy of vaporization per cubic centimeter, the CED, is calculated by subtracting the product of the gas constant and the temperature from ΔH_v and dividing by the molar volume. Similar values on nonvolatile materials such as polymers cannot be determined directly. They are obtained from maxima in plots of solvent absorption versus CED. The CED range is obtained from use of a series of solvents covering this range. The polymers must be lightly crosslinked. CED values may also be approximated from viscosity measurements or by summation of the molar attraction constants assigned to the chemical groups in molecules.

Applications

1. The search for the best solvent for a polymer is greatly aided once the cohesive energy density of the polymer is known. An example of this is to find the solubility parameter of a cross-linked copolymer of styrene and divinyl-benzene by measuring its swelling in different solvents of known solubility parameters. A function of the swelling ratio Q and the molar volume V_1 of each solvent is plotted against the solubility parameter in Figure 1. The value of the solubility parameter that makes the swelling function go to zero is the solubility parameter of the polymer. Solvents with a solubility parameter of 9.0 will most likely be the best solvents for this copolymer.
2. Alternately, knowledge of the solubility parameter aids in the selection of a solvent to precipitate the polymer from solution.
3. Plasticizer compatibility in a polymer can be predicted by a comparison of the CED's.
4. CED information is useful in the selection of cleaning solvents which will not dissolve plastic objects.

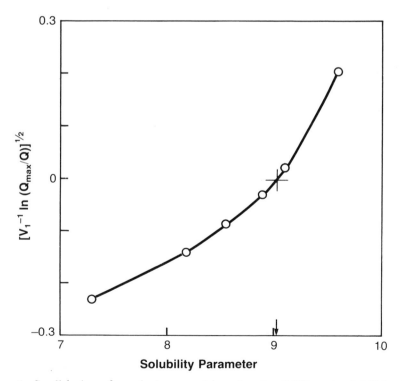

Figure 1. Swell index of a polystyrene gel in solvents of different solubilities indicating the polymer solubility parameter.

5. CED has been correlated to critical surface tension of a bulk polymer, ultimate material strength, internal pressure, and compressibility; however, it is often easier to measure these other quantities directly than the CED.

Limitations

Cohesive energy density is a single parameter which attempts to incorporate a complex set of solvent-solute interactions: dispersion (or London), dipole-dipole, and dipole-induced dipole forces. Hydrogen bonds constitute a special type of dipole-dipole interaction. Since the CED gives an interaction parameter reflecting all these effects, it is possible that two materials may have similar solubility parameters for very different reasons. In that case, they may indeed be incompatible despite the equality of their solubility parameters. More complex schemes that account for different interactions separately have been developed but are more burdensome to use.

References

(1) Barton, A. F. M., CRC Handbook of Solubility Parameters and Other Cohesion Parameters; CRC Press: Boca Raton, FL, 1983.

(2) Kaelble, D. H., Physical Chemistry of Adhesion; John Wiley and Sons: New York, 1971; pp. 45–116.

(3) Morawetz, H., Macromolecules in Solution; John Wiley and Sons: New York, 1965; pp. 33–90.

Acknowledgment

Data in Figure 1 is reprinted from: Boyer, R. F.; Spencer, R. S., J. Polym. Sci., 3, 1948, 108; copyright© 1948; with permission of John Wiley and Sons Inc.

SURFACE ENERGY OF SOLIDS

Use

Surface energy measurements on solids are used to obtain information on the chemical nature of the groups exposed at the surface. Such information is important in development of more efficient lubricants and adhesives and in evaluating wetting and soil-resistance characteristics of fabrics.

Sample

Solid samples in flat-surface or filament forms are required. Their surfaces should be as smooth as can be obtained. Flat-surface solids as small as one square inch are sufficient for measurements.

Principle

Surface energy values of solids in flat-film and filament forms are obtained by measuring the contact angles between different liquids and the solid surfaces. A linear plot is made of the cosines of the contact angles measured versus the surface tensions of the liquids used. A straight line drawn through the data points is extrapolated to cosine = 1.0 (zero angle). The surface tension corresponding to the point at which this line intersects cosine = 1.0 is defined as the surface energy of the solid being measured.

Contact angles are measured in flat films from direct observation of a liquid drop. The angle formed between the liquid-vapor and the solid-liquid interfaces is determined with an optical contact angle goniometer. For a fiber or a plate of constant cross section, a Wilhelmy balance like that shown in Figure 1 gives the most accurate measurements. Here a microbalance measures the force needed to insert or withdraw the solid at a constant rate from a liquid. The force recorded by the balance is the apparent weight of

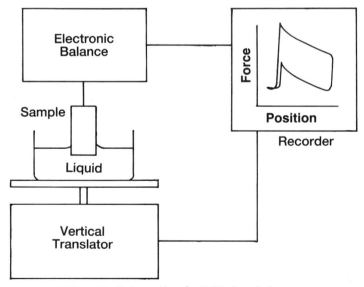

Figure 1. Schematic of a Wilhelmy balance.

the object. At the instant of contact with the liquid, the force increases since the surface tension of the liquid pulls the object into the liquid. With deeper insertion, the force then decreases because the buoyant force of the submerged object acts to counter the weight and the surface force. The change in force on insertion, the surface tension of the liquid, and the perimeter of the object determine the contact angle for the solid-liquid pair. On many surfaces the contact angle on insertion (advancing) and withdrawal (receding) are not equal. The values may also show variation with subsequent measurements.

Applications

1. Surface energy measurements can be used to assess changes in the surface of a film or fiber by the addition of an ingredient or a variation in the processing. Figure 2 compares the surface energy of a pure nylon monofilament to a similar monofilament impregnated with an additive. The decreased surface energy of the treated fiber indicated enhanced soil resistance.
2. Surface cleanliness can be measured since dirt on the surface will change the surface energy.
3. The stability of emulsions is predicted by changes in the surface tension of liquids upon the addition of surfactants.
4. Surface adhesion of paints, inks, and adhesives is determined by surface energies.

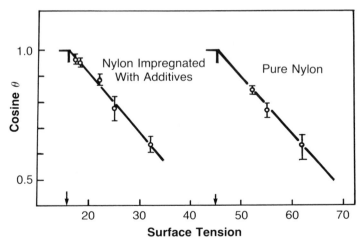

Figure 2. Extrapolation of contact angle data to obtain surface energy of pure and modified nylon monofilaments.

Limitations

Meaningful measurements of contact angle on a film require surfaces that are uniform in composition, cleanliness and roughness. When measurements are made on a fiber, the material must also have uniform cross sections with known perimeter. If one is seeking surface energy density of the material, it is necessary for the surface to be smooth. The amount of hydrogen bonding in the test liquids somewhat influences the measured surface energy.

References

(1) Adamson, A. W., Physical Chemistry of Surfaces; John Wiley and Sons: New York, 1982; pp 332–368.
(2) Gould, R. F., ed., Contact Angle, Wettability, and Adhesion; American Chemical Society: Washington, D.C., 1964.

DILATOMETRY

Use

Dilatometric techniques can be used to measure cubical coefficients of thermal expansion, first and second order transition temperatures, and rates of crystallization. Although other techniques (see subsections (1) Differential Thermal Analysis and Differential Scanning Colorimetry and (2) Thermomechanical Analysis and Dilatometric Analysis) can sometimes provide this information more readily, dilatometry offers certain advantages in measurement of the cubical coefficient of expansion of molten polymers.

Sample

Approximately 20 g of solid sample are required. Analysis can be done on smaller samples, but with some sacrifice in accuracy.

Principle

The specific volume (cm^3/gm) is a function of temperature because molecular motion changes with temperature. Measurements are made to determine the specific volume over the temperature range of interest, and the data is plotted as a function of temperature. In temperature regions in which no phase transitions occur, the curve is smooth, and the slope gives the cubical coefficient of expansion at that temperature. A discontinuity or an abrupt change in slope in this curve indicates a first order or second order phase change, T_m and T_g, respectively.

Precision

Dilatometric measurements can provide the temperatures of melt and glass transitions to a precision of $\sim 1\%$. The actual temperatures obtained from these transitions depend on equilibration time allowed at each temperature; consequently, equilibration time is usually chosen as that primarily experienced in end processing conditions.

Applications

1. Change in density or volume of a material with temperature can be calculated with the coefficient of thermal expansion.

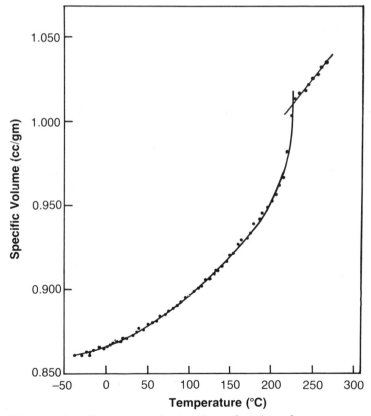

Figure 1. Specific volume of nylon 6 as a function of temperature.

2. Melt temperatures of polymers are observed by a discontinuity in the coefficient of thermal expansion. As an example, the plot of the specific volume of nylon 6 between $-50°$ and $225°C$ is shown in Figure 1. The melting point at $225°C$ is evident from the cusp. Just below this temperature the thermal coefficient of cubical expansion is a strong function of temperature. Above this temperature the thermal coefficient of the melt is a characteristic property of the resin being evaluated.
3. Glass transitions in polymers produce continuous changes in the coefficients of thermal expansion.
4. The temperature for various transitions are indicative of the polymer, its additives, and the degree of crystallinity.

Limitations

Accurate measurements require that the sample fill the dilatometer as fully as possible. Thus, pellets and spheres which give large interparticle spacing are not recommended. Remolding these forms into more compact, stacking plaques improves the accuracy but may change the density and the coefficient of volumetric expansion of the material to be characterized.

References

(1) Bauer, N.; Levin, S. Z., Physical Methods of Organic Chemistry; Part I, Weissberger, A., ed., Interscience Publishers Inc.: New York, 1959; pp 131–190.
(2) Bekkedahl, N., J. Research Natl. Bur. Standards, 43, 1945, 145.

CHAPTER 12

Physical Testing

Igor Palley, Anthony J. Signorelli

MECHANICAL PROPERTIES OF MATERIALS

Use

Mechanical testing methods are used to evaluate materials under a variety of loading conditions. Mechanical properties that can be evaluated include: moduli (characterizing their rigidity); ultimate or other characteristic values of stresses and strains; and specific work or energies characterizing their strength, ductility, resilience, and general toughness, respectively.

Sample

The sample must be a solid material. Sample dimensions depend on the method of characterization and the material's microstructure. The sample should be large enough to represent the structural part and to allow evaluation by continuum mechanics. A typical sample may be in the shape of a bar or a rod. Samples can also be in the form of fibers, films and sheets.

Principle

Continuum mechanics provides mechanical testing with an analytical tool for understanding and characterizing a material's behavior and performance. For engineering materials the infinite number of loading situations can be rationalized by continuum mechanics and interpreted in terms of stress/strain states. The rigidity of materials can be expressed in terms of moduli (the matrix of rigidity) which can be found from a limited number of properly designed tests which measure stress/strain relationships.

The strength of materials is expressed by the ultimate values of stresses (in general by the stress tensor) which can be found by loading the sample until failure. For example, loading can be in tension, compression or shear. Other loading situations are desirable because of simplicity of loading and interpretation of their stress state, e.g., flexing. In order to reduce the effects of certain factors that are not accounted for in the employed models of mechanics (effect of microstructure, scale effects, stress concentration around the loading areas) the sample size and loading conditions are standardized by ASTM, Military and Federal specifications. For all cases, stress-strain diagrams are analyzed and the moduli and the characteristic stresses are reported. A typical stress-strain curve for a plastic material is shown in Figure 1.

Applications

Mechanical properties can be determined for all classes of materials: metals, ceramics, polymers and composites. Testing standards can vary for different classes. The measured properties are utilized in materials research and development, quality control, and in structural design and evaluation.

Mechanical properties can be evaluated under a wide variety of temperatures, strain rates and environmental conditions. In particular, one can evaluate mechanical properties in the temperature range −150°C to +1700°C. The laboratory environment should be maintained at a constant temperature of 23°C and 50% RH. The modern laboratory has computer-

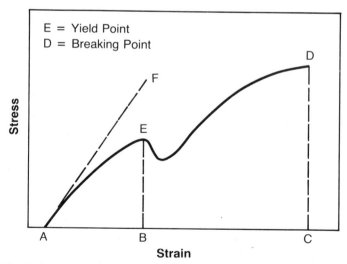

Figure 1. Typical stress-strain curve for plastic material showing: breaking strength, DC; elongation at break, AC; yield strength, EB; yield elongation, AB; work to break, area ACDA; work to yield, area AEBA; Young's modulus, slope of line AF.

ized universal constant displacement rate testers with accessories for all types of standardized testing for all classes of materials. When needed, a large variety of other mechanical tests for new materials, structures, and loading conditions can also be designed and performed. For example, special equipment for composite adhesion studies and for compression strength after impact can be designed.

A well equipped laboratory also has a variety of standard mechanical devices designed for particular types of tests. Examples are:

- Non-instrumented impact testers: Izod, Charpy, Gardener, Tensile Impact and Falling Ball. (See section on Instrumented Impact Tester.) In Figures 2 and 3 are shown schematics of an Izod and Charpy Tester.
- Hardness measurement instruments.
- Four point Heat Deflection and Vicat Softening Point Tester. These techniques are shown in Figures 4 and 5.

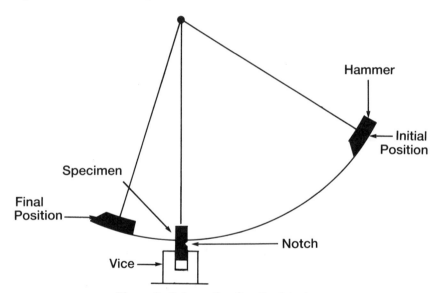

Figure 2. Schematic of an Izod tester.

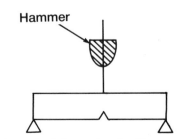

Figure 3. Schematic of a Charpy tester.

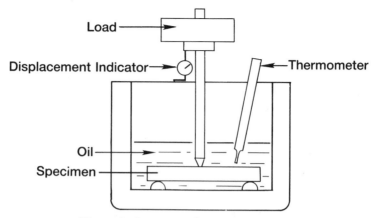

Figure 4. Apparatus for heat deflection.

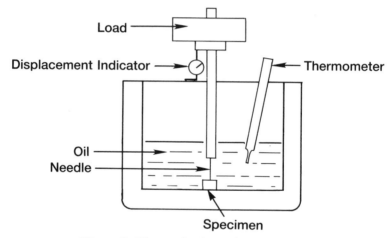

Figure 5. Vicat softening point apparatus.

- A variety of abrasion testing machines—Tabor, Wyzenbeck, Inflated Diaphragm, Mar Resistance (Falling Grit). Two of these are shown in Figure 6.
- Dead-load material creep evaluation devices.
- Torsional Modulus instrumentation.
- Thermal Conductivity equipment.
- Density gradient columns.
- Graves and Elmendorf Tear Testers.
- Coefficient of Friction equipment. In Figure 7 is given a schematic of a coefficient of friction apparatus.
- Haze and gloss analyzers.

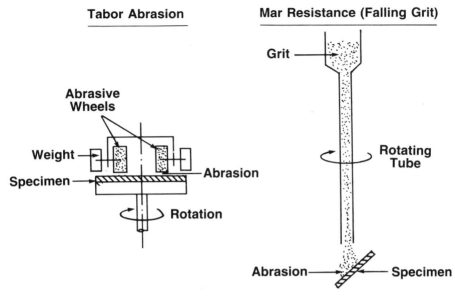

Figure 6. Schematics of abrasion testing machines.

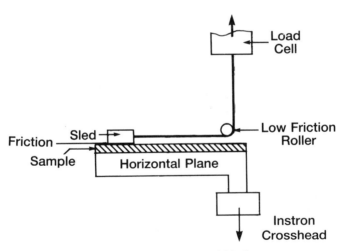

Figure 7. Schematic of coefficient of friction apparatus.

A well equipped laboratory should also have material and sample fabrication capabilities. Equipment to mix or blend material components, as well as mold, cure, sheet-out, and extrude polymers is also useful.

Limitations

Since mechanical properties of materials are determined by using specific simple tests, like uniaxial loading, and a specific size of the specimen, the information should be used with caution whenever the loading conditions and dimensions of the real part are different from those of the specimen.

References

(1) Brostow, W.; Corneliussen, R. D., Failure of Plastics; Hanser Publishers: Munich-Vienna-New York, 1986.
(2) Hartmann, F., The Mathematical Foundation of Structural Mechanics; Springer-Verlag, 1985.
(3) Sih, G. C.; Skudra, A. M., Failure Mechanics of Composites; Elesevier Science Pub.: Amsterdam; New York, 1985.
(4) Timoshenko, S. P.; Gere, J. M., Mechanics of Materials; Van Nostrand Reinhold Co.: New York, 1972.
(5) Whitney, J. M.; Daniel, I. M.; Pipes, R. B., Experimental Mechanics of Fiber Reinforced Composite Materials; SESA, 1982.

FATIGUE TESTING

Use

In fatigue analysis a material is subjected to periodically fluctuating loads. The load is designed to simulate the conditions that the material will ultimately experience. Thus, fatigue analysis can be used to determine whether a specific material will serve without failure for a given number of loading cycles.

Samples

Samples are blocks of solid materials of sufficient dimensions to be representative of the structural components and to be justifiably analyzed with the methods of continuum mechanics. In general, the samples have the shape of elongated cylinders (of round or rectangular cross-section) with a smooth (unnotched) surface or a sharp notch.

Principle

Depending on material applications, the samples are tested either unnotched or with sharp notches. The latter provide a high level of stress concentration. Tests using unnotched samples provide information related to crack initiation. Testing samples with sharp notches relies on the concepts of fracture mechanics which were originally developed for static loading conditions. The use of notched test bars leads to a more basic approach to the analysis of fatigue failure mechanisms and is closely related to crack propagation. Loading can be in tension and/or compression, torsion, or bending. In practice, the loading conditions are chosen in such a manner as to clearly approximate the conditions to which a particular structural component is subjected. A typical cycling stress controlled loading program (Figure 1) can be characterized by constant values of two parameters σ_{max} and σ_{min} (or σ_a and σ_m, or $R = \sigma_{min}/\sigma_{max}$ and σ_a). Usually the shape of the cycle is sinusoidal (Figure 1), but many other shapes can be used.

Applications

Fatigue analysis can be used to study the deformation and fracture responses of metals, engineering plastics, fibers, ceramics and all types of composite materials when they are subjected to cyclic loads. The data obtained can be used for basic studies, material design and development, material selection, prediction of component life, and for quality control.

In addition to cycling stress-controlled loading one can also carry out another type of cycling loading program (called strain-controlled loading) which maintains the maximum and the minimum values of strain (ϵ_{max}, ϵ_{min}) constant. Figure 2 is an example of a cycling stress–strain diagram (hysteresis loops) for a strain hardening material. Figure 3 shows a typical relationship between σ_a for stress controlled testing (R-const) and N, the number of cycles at which the failure takes place. Such diagrams are the basis for fatigue analysis using unnotched samples.

Figure 4 shows a typical relationship between da/dN—the crack length increment for a cycle (where a is the crack length and N is the number of

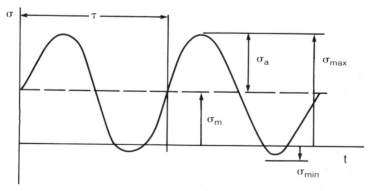

Figure 1. Typical cyclic stress controlled loading.

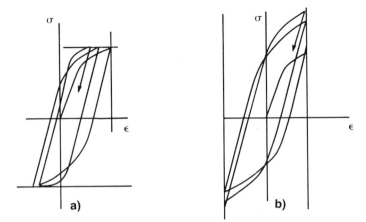

Figure 2. Hysteresis loops for (a) stress controlled and (b) strain controlled cyclic loading strain hardening material (low cycle fatigue).

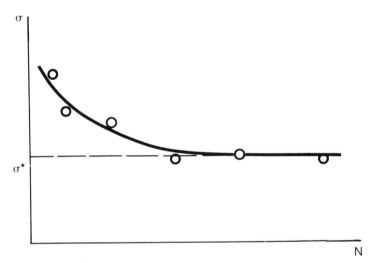

Figure 3. Typical fatigue responses in metals. σ^* = endurance limit. Unnotched sample.

cycles), and the difference ΔK between the extremes of the stress intensity factor, K. These types of diagrams represent the basis for fatigue analysis in experiments with notched samples.

For basic studies and material design, mechanical fatigue analysis is especially effective in conjunction with macrofractography (optical microscopy), microfractography (SEM), and analysis of the crack fracture surfaces and the damaged zone surrounding the tip and the path of the crack. Other auxiliary techniques might include: analysis of the decrease in static strength (determination of residual strength) and the modulus of materials subjected to a number of loading cycles.

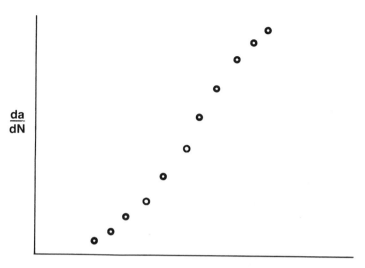

Figure 4. Typical relationship between the rate of crack propagation and ΔK for polymers. Notched sample.

The following problems common to all classes of materials can be evaluated in fatigue analysis:

- Effect of stress-strain history of loading including the current state.
- Effect of thermal history.
- Effect of temperature.
- Effect of environment.
- Scale effect.

Limitations

Only in very brittle materials fatigue damage develops as a single major crack, propagating from the notch. More usual is the development of a multitude of microcracks around the notch (or the major crack tip) before the major crack moves further through the already damaged material. More complex interpretations can be used in these cases. For example, the concept of a 'crack layer' can be invoked.

References

(1) Fong, J. T., ed.; Fatigue Mechanisms; ASTM; STP 675, 1979.
(2) Hertzberg, R. W.; Manson, J. A., Fatigue of Engineering Plastics; Academic Press: New York, 1980.

(3) Lauraitis, K. N., ed.; Fatigue of Fibrous Composite Materials; ASTM: STP 723, 1979.

(4) Ritchie, R. O.; Lankford, S., ed.; Small Fatigue Cracks; Metallurgical Soc: 1986.

INSTRUMENTED IMPACT ANALYSIS

Use

Instrumented impact analysis is the basis for determining the relationship between the dynamic force and displacement during the process of impact loading. Through the measurement of force, displacement, energy and velocity one can obtain important information that is related to the amount of energy a material can absorb during plastic deformation.

Sample

The sample can be in the form of molded plaques (typically 4 inches square and 0.125 inches thick), bars, or an actual structural component.

Principle

By definition, impact is the striking of one body against another. Collisions at high accelerations result in immediate changes of velocities and, therefore, strong forces of interaction which cause materials to fail. The instrumented impact tester is provided with a means to record and automatically display, in real time, the dynamic force on the sample as well as, the corresponding velocity and displacement under the force trace. From the force-displacement diagram (Figure 1) the rigidity of material response (dynamic modulus), the maximal force and the energy dissipated by the sample at different stages of the loading process are established.

A schematic of an instrumented high speed puncture test is given in Figure 1.

Applications

Impact analysis and instrumented impact testing can be used as an R&D tool, a technology or quality control device, or as a means of establishing materials design parameters. Technology, dealing with transportation (aerospace, automotive, etc.), protection (civil engineering structures, armor, personnel protection, packaging), sporting goods, and communication, are the

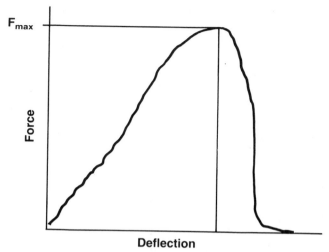

Figure 1. Typical relationship between the force and the deflection under the force recorded by an instrumented impact tester.

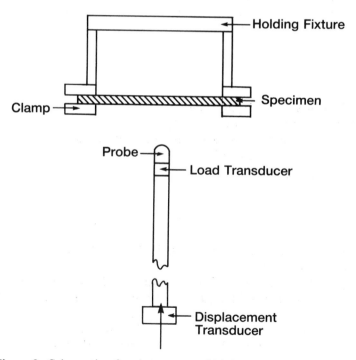

Figure 2. Schematic of an instrumented high speed puncture tester.

prime users of impact analysis. Material parameters such as impact strength, damage tolerance, resistance to perforation, and dynamic response can be obtained.

Specifically, *impact strength* is established by observation of the destruction of a material sample or a structural component. *Damage tolerance* indicates the ability of a component to experience an impact without significant structural changes or with a minimal effect on its designed load bearing capabilities. *Resistance to perforation* is an important characteristic for some load bearing structures having the additional function of protecting personnel and equipment from the perforating body or the surrounding media (gas, liquid). Two aspects of *Dynamic Response* of structure to impact can be studied by using instrumented impact. First, for loads below those causing any damage, the force to deflection relationship (particularly, the dynamic modulus) and the damping characteristics of the component can be established. Secondly, energy absorption capability through plastic deformation of the structure can be found. Instrumented impact also allows one to study crack propagation problems (see Fracture Analysis section) under conditions of high loading rate.

Finally, it should be noted that there are no restrictions on the shape of the impacted object. Using an environmental chamber, materials can be studied under conditions of temperature that range from $-150°C$ to $+500°C$. Impact tests can be carried out at velocities up to 21 m/sec (50,000 inches/min or ~47 miles/hour) and with an impact energy up to 1800 J.

Limitations

A problem of separating the significant signal from background noise is always present in the Instrumented Impact Test because of the fundamental nature of the impact process itself. Filtering is often desirable to clarify the data, but should be applied with caution and understanding to avoid loss or distortion of the significant information.

References

(1) Brostow, W.; Corneliussen, R. D., Failure of Plastics; Hanser Publishers: Munich-Vienna-New York, 1986.

(2) Kessler, S. L.; Adams, G. C.; Driscoll, S. B.; Ireland, D. R. Instrumented Impact Testing of Plastics and Composite Materials; ASTM: STP 936, 1986.

(3) Zukas, J. A.; Nicholas, T.; Swift, H. F.; Greszczuk, L. B.; Curran, D. R., Impact Dynamics; John Wiley & Sons: New York, 1982.

FRACTURE TOUGHNESS OF MATERIALS

Use

Fracture toughness is an important engineering characteristic of solid materials. Fracture toughness is a measure of the degree of resistance of a material when under a load, to its sudden catastrophic failure through crack propagation. This characteristic can be determined for metals, ceramics, plastics, and composites of all types (polymer, cementitious, ceramic and metal).

Sample

The sample must be a solid. Sample dimensions depend on the applied method and the material structure and characteristics (the coarseness of the microstructure, anisotropy and plasticity). For example, a short bar fracture toughness method allows the use of samples with cross-sectional dimensions of 12.5 mm on a side and 19 mm long. The thickness of fiber reinforced composite samples characterized by the double cantelever beam method can be as small as 3 mm. Larger samples can be tested when the material is available and preparation of the larger samples does not represent a problem.

Principle

The theory of crack propagation for brittle or relatively brittle materials uses the idealized model of an infinitely sharp notch. For the idealized model, which is the basis for Linear Elastic Fracture Mechanics (LEFM), stresses are infinitely high. (In reality, this can not occur because the plastic zone at the front of an advancing crack tip prevents this). To overcome this problem, LEFM uses as a criterion the concept of a stress intensity factor. The stress intensity factor (K) is a coefficient in the expression for stress distribution around an infinitely sharp notch. It is a function of load, sample and the notch size. As a criterion for crack propagation LEFM uses the critical value of the stress intensity factor (K_{1c}) which is also called fracture toughness.

The stress intensity factor has a one-to-one relationship with a function called the energy release rate (G). In terms of this model the crack will propagate when the energy release rate is equal to a certain critical value (G_{1c}). Both approaches, the one based on force (K_{1c}) and the one based on energy (G_{1c}), can be used to obtain fracture toughness.

When materials show significant nonlinear behavior prior to crack formation (high toughness material), the LEFM approach is not valid, because the plastic or nonlinear deformation zone around the crack tip is too large. For these types of samples a non-LEFM approach is used. This approach is based on the concept of J-integral or a certain path-independent integral function of stresses, and its critical value J_{1c}.

Applications

Fracture toughness values can be readily obtained for samples of engineering materials (ceramics, metals, polymers and composites). A completely automated, computer controlled test system, utilizing small short rod/short bar samples (Figure 1) provides fracture toughness information for materials with fracture toughness up to 269 MPa $\sqrt{m}$ (240 psi $\sqrt{in}$) using both the LEFM and non-LEFM approach. Fracture toughness (Figure 2) can be related to materials structure and processing conditions and can be used with fracture surface and damaged area observations and analysis. This information can suggest ways to improve the fracture toughness of a material.

In simple cases, fracture toughness is reported as a number that is characteristic of a material's resistance to crack propagation. However, in many cases, fracture toughness may be a function of the crack propagation process (Fracture Toughness Resistance Curve). Such behavior can be analyzed in the laboratory.

Conventionally, fracture toughness is determined in the process of quasistatic loading (Figure 3). Fatigue and impact failure of engineering materials are also accompanied by crack propagation. Analysis of fatigue and impact properties using fracture mechanics can be provided.

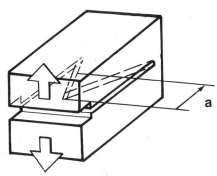

Figure 1. Short bar chevron notched (SBCN) testing specimen. Arrows indicate applied load.

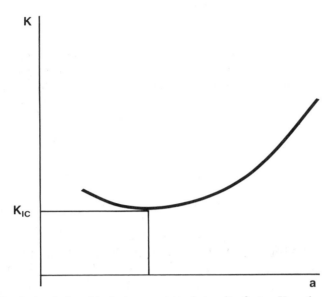

Figure 2. Typical relationship between stress intensity factor K and crack length **a** for SBCN specimen.

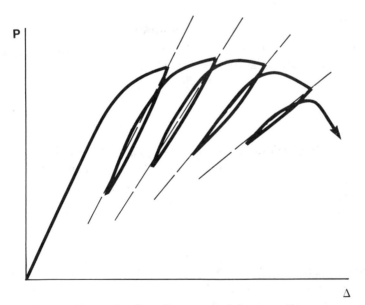

Figure 3. Typical loading/unloading diagram used for compliance or energy dissipation methods in fracture analysis.

Limitations

The basic concepts used to describe fracture phenomena are derived from the principles of continuum mechanics. As such the concepts are applicable if and only if the structural element being studied is representative of the continuum.

References

(1) Evans, A. G., ed.; Fracture in Ceramic Materials; Noyes Publications: New York, 1984.

(2) Kobayashi, A. S., Experimental Techniques in Fracture Mechanics; SESA: 1973.

(3) Liebowitz, H. ed.; Fracture; Academic Press; New York, 1968.

(4) Underwood, J. H.; Chait, R.; Smith, C. W.; Wilhelm, D. P.; Andrews, W. A.; Newman, J. C., ed.; Fracture Mechanics; ASTM: STP 905, 1984.

(5) Underwood, J. H.; Frieman, S. W.; Baratta, F. I., Chevron-Notched Specimens, Testing and Stress Analysis; ASTM: STP 855, 1987.

GAS AND LIQUID PERMEABILITY

Use

The technique measures the transmission rate of a diffusing substance through polymer films, laminates, coated papers, and flexible or semi-rigid packaging materials.

Sample

Flat barrier films should have an area of at least 10 cm² for oxygen, liquid, and water vapor permeability measurements. Other examples of specimens that are suitable for permeability testing are bottles, pouches, and thermo-formed cups.

Principle

Permeability measurements characterize those processes in which a quantity of liquid or gas is transported from one side of a barrier film to another. Suppose a constant concentration of a diffusant gas or liquid is maintained on one side of a barrier film of thickness, d. The presence of a concentration gradient results in a transfer of diffusant from one side of the film to the other. The diffusant concentration, c, in the barrier film at a distance, x, from the surface as a function of time, t, is given by Fick's Second Law:

$$\frac{\partial C}{\partial t} = D \frac{\partial^2 C}{\partial X^2}$$

where D is a constant referred to as the diffusion coefficient. Fick's Second Law yields the concentration of diffusant as a function of x and t. Under steady state conditions, i.e., $\partial C / \partial t = 0$, the concentration of diffusant which passes through the barrier does not change with time. From a plot of the concentration of diffusant which passes through the film versus t, one can determine the permeability of the film for a given diffusant at a specified temperature and relative humidity. The units for permeability are concentration per unit area per unit time, e.g., cc/100 in^2/24 hours.

Applications

Permeability measurements play an extremely important role in the packaging industry. The barrier properties of a material can be used to determine if a particular film, either as a single layer or as part of a laminate, will be suitable for packaging foods, electronic materials/devices, pharmaceuticals, etc.

In general, in a system that is separated from the surroundings by a barrier film, permeability measurements involve monitoring the weight gained or lost by the system. For example, liquid permeability measurements can be made using Bixter-Michaels cups. In this method, one monitors the weight loss of liquid through the barrier film to be tested vs. time. Water vapor transmission rates for materials in sheet form are determined using ASTM E96. The tested film is used as a barrier and is sealed to a dish containing a desiccant. The system is placed in a chamber maintained at 95% RH and 100°F. Moisture gain, as a result of water vapor penetrating the film, is determined by following the increase in the weight of the desiccant.

Oxygen transmission rate (OTR) through a flat barrier film can be determined using a commercial oxygen permeability unit. The experimental set-up consists of clamping a film in a horizontal position in a diffusion cell and allowing O_2 to flow over the top surface of the film. A carrier gas, consisting of N_2/H_2, flows over the bottom surface of the film. The carrier gas is first passed through a catalyst bed in order to remove residual O_2. After a period of time (referred to as the conditioning time) the amount of O_2 permeating through the film and entering the carrier gas reaches a constant value. As the carrier gas picks-up the O_2 and flows through the sensing cell a current is produced. The magnitude of this current, in accordance with Faraday's Law, is related to the oxygen concentration that has permeated through the barrier film, i.e., OTR. Typically, one can measure OTR at 0% RH as a func-

TABLE 1. REPRESENTATIVE OTR DATA 23°C
(cc/mil/100 in²/24 hr.)

Sample	%RH	OTR
Nylon 6	0	2.8
Nylon 6	100	14.2
Nylon 6/6,6	0	2.0
Nylon 6/6,6	100	15.0

tion of temperature. One can also measure OTR as a function of RH for moisture-sensitive barrier films. Table 1 shows some representative OTR data for unoriented nylon 6 and nylon 6/6,6 films as a function of relative humidity (constant temperature). Temperature also plays an important role in the rate of oxygen transmission. For example, between 21°C and 35°C (0% RH) the measured OTR value for nylon 6 doubles.

Limitations

In package design, the effects of temperature and humidity on the gas and/ or liquid transmission rate are extremely important. For some polymers, these measurements can be complicated by the following: (1) morphological changes in the polymer film brought about by temperature, moisture, or organic vapor; (2) micro-crazing or stress corrosion cracking of the film; (3) permeability measurements that follow weight gain or loss via weighing subject the film to stress as the sample is cycled from chamber temperature to ambient and back to the chamber; (4) in general, the measurements are time-consuming.

References

(1) Crank, J.; Park, G. C., Diffusion in Polymers; Academic Press: New York, 1968.

(2) VanKrevelen, D. W., Properties of Polymers; Chapter 18, Elsevier: New York, 1976.

CHAPTER 13

Scientific Computation

Sheldon Eichenbaum, Kamal Sarkar, Daniel H. Stevans, David Y. Hsieh, Gregory J. Czerwienski, Steven E. Sund, Eli Rosenthal, Willis B. Hammond

APPLIED FINITE ELEMENT ANALYSIS

Use

The Finite Element Method is used to study:

A. deformation and stress data in structural analysis;
B. temperature distribution and heat flow in thermal analysis;
C. coupled thermomechanical interaction data;
D. materials interaction and fracture data;
E. velocity vector and flow field in fluid dynamics;
F. electrostatic and magnetostatic field vectors for electromagnetic analyses.

Principle

1. The Finite Element Method is a numerical technique which approximates a solution for energy distribution in a given physical problem, and forces the approximate solution toward the exact solution by using variational (energy minimization) principles.
2. The method is efficiently implemented in a variety of algorithms in computer programs, but usually involves the following steps:
   - the definition of an initial physical configuration and boundary conditions (preprocessing);
   - matrix operations to solve the system of equations (the run);

- graphical display of the resulting changes in the system (postprocessing).

a. Preprocessing involves the specification of spatial coordinates for a mesh of points (nodes) and their interconnections. Nodal data represent the relationships between points in a physical object on a surface (2-D) or inside (3-D) the model. The data also specify physical properties at each point, line, surface or solid. Elements, which are areas or volumes enclosed by nodal boundaries, are treated as homogeneous or composite pieces of material. There are hundreds of types of basic element shapes, and each type has a special algorithm for solving a particular kind of physical problem efficiently.

b. The run requires the selection of criteria which govern the choice of algorithm, the speed and accuracy of the solution, and certain limits which control convergence of iterative processes and matrix operations. The correct formulation of the matrix of equations for the physical model and an appropriate choice of method for solving this matrix, will determine whether the results will closely simulate a real physical situation.

c. The selection of the method of postprocessing is often crucial to the correct interpretation of the volumes of coordinate and property data generated during the iterations of the algorithm. Usually the data is presented in contour plots and displacement diagrams. Multicolor graphics is one very efficient way to study data variations and to highlight selected conditions of stress, temperature, field intensity, and so forth.

Applications

The wide usage of this method over the last couple of decades, in particular, has demonstrated its applicability in almost every major field of engineering discipline. Moreover, the popularity of this method has made it extremely user-friendly. Apart from the universality of its applicability, the elegance of the method is in its ability to handle extremely complex geometries. Some examples of current applications are as follows:

1. Prediction of deformations and stresses in two or three-dimensional structures;
2. Prediction of temperatures and heat flow rates in structural components;
3. Prediction of thermal and structural response of a system subjected to coupled thermomechanical loading;
4. Identification of an optimizing material for a particular structural performance under a given set of thermomechanical loading;
5. Prediction of velocity profiles, flow rates and pressure drop of fluid across a complex geometry;

6. Simulation of special material behaviors like composites, fracture and failure;
7. Prediction of electrical and magnetic field lines in complex conductive material.

In the specific application cited in 3, the temperature profile developed in a fully loaded (5000 lbs/tire) truck tire running at 65 MPH was predicted by modeling the quasi-steady, fully coupled thermomechanical problem (Figure 1) with a finite element mesh. The most interesting aspect of the fully developed temperature profile was the multiplicity of the "hot spots" in the tire. The temperature in these regions is as high as 130°C. This unknown fact was verified later experimentally with the help of the model.

Limitations

1. If and when the nature of the governing differential equations change, this method cannot be utilized (e.g., transonic flow problems).

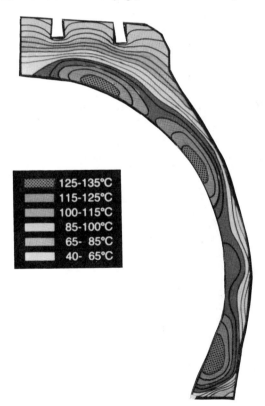

Figure 1. Computer prediction of the thermomechanical behavior of pneumatic tires.

2. If the nature of the governing partial differential equations are not known in enough detail to use an extremum and/or stationary (energy) principle, the method cannot be utilized (e.g., powder consolidation/compaction problems).
3. Many real-life problems have too many variables to be modeled in a cost/effective manner (e.g., ballistic impact on a laminated composite plate).

References

(1) Desai, C. S.; Abel, J. F., Introduction to Finite Element Method, A Numerical Method for Engineering Analysis; Van Nostrand Reinhold Company: New York, 1972.

(2) Rao, S. S., Finite Element Method in Engineering; Pergamon Publishers: New York, 1982.

(3) Zienkiewicz, O. C., The Finite Element Method; Third edition, McGraw-Hill Publishers: New York, 1983.

COMPUTATIONAL FLUID DYNAMICS MODELING

Use

Computational Fluid Dynamics Modeling is used to determine the detailed flow characteristics of fluids flowing in, around and through various types of devices, equipment and apparatus. The modeling may include the consideration of, and the effects of, diffusion, heat transfer, chemical reactions and multiphase flows. The results of the modeling are presented numerically, graphically or descriptively, and include any or all of the following:

a) pressure contours
b) flow patterns
c) velocity profiles
d) turbulence conditions
e) convective heat transfer and temperature distributions
f) dispersed phase concentration distributions
g) chemical reaction profiles

Principle

Computational Fluid Dynamics Modeling uses computer based numerical methods for solving the differential and auxiliary equations used to describe fluid flow. Fluid flow has been described by a set of partial differential equations known as the Navier-Stokes equations. In addition, turbulence, heat transfer and chemical reactions require additional partial differential equations to describe their effects. The physical properties of the fluids, as functions of a set of independent variables such as temperature and pressure, along with the flow geometry and the boundary and initial conditions of the flow are needed to complete the definition of the physical system.

The different numerical methods used to solve the equations are known as finite difference, finite volume or finite element methods. The three steps used to get the results of interest are:

1. Pre-processing: The geometry is divided into small regular domains interconnected through nodes. The physical properties of the fluid are given in these domains as a function of temperature, flow rate, etc. Boundary and initial conditions are also specified to complete the description of the problem. In general, this part of the modeling is highly interactive and graphical, in nature.
2. Solution: The solution for the velocity vectors, pressure contours, turbulence generation, second phase distribution, etc., is done in this segment of the modeling. In general, an iterative procedure is used to reach a certain level of convergence for each of these parameters. These tolerances are either built-in or user-specified.
3. Post-processing: Once a converged solution has been reached, the results are displayed as contours/graphs on any given plane/surface of the geometry. In general, this part of the modeling is highly graphical in nature and demands good insight into the problem.

There are several commercial computer codes available which have been developed to solve the fluid flow equations.

Applications

The use of commercial computer codes has made the realistic use of computational fluid dynamics possible. Some of its uses are:

1. Calculations of flow patterns, pressure contours, turbulence conditions, temperature and heat flow in complex three-dimensional geometries such as pumps, turbines, heat exchangers, etc.
2. Identification, distribution and interaction of a second phase in a multiphase flow problem, e.g., dispersion of a pollutant.
3. Flow patterns, pressure contours and dust distribution in a flow field containing a filter element. (See Figure 1 for an example of a pressure contour plot of dust laden air in a heavy duty air filter assembly).
4. Second phase distribution, temperature profile and flow field in a reacting multiphase fluid flow problem in a combustion chamber.

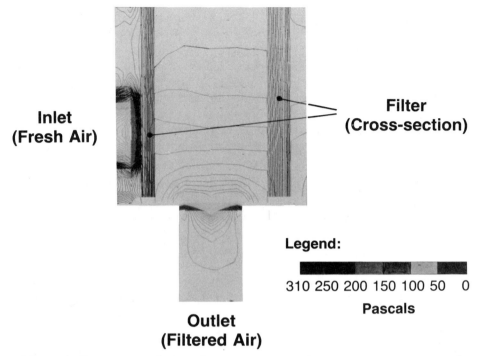

Inlet (Fresh Air)

Filter (Cross-section)

Legend:

310 250 200 150 100 50 0

Pascals

Outlet (Filtered Air)

Figure 1. Computer prediction of pressure contours for flow through an air filter.

The following industries have made extensive use of computational fluid dynamics modeling:

1. Chemical and Metallurgical industries (furnace, heat exchanger, condenser, reactor, etc.)
2. Electrical and Electronic industries (heat flow through power generator, electronic components, etc.)
3. Aerospace industry (heat, mass and fluid flow in and around planes, rockets, etc.)
4. Automotive industry (chemical combustion, air filtration, flow and turbulence around moving cars, etc.)
5. Biomedical Engineering (heat and mass flow in human body)
6. Environmental Engineering (weather prediction, pollution control, etc.)
7. Material Processing (fiber drawing, fiber spinning, etc.)

Limitations

The accuracy of the predicted results of fluid dynamics modeling is strongly dependent on adequacy of the mathematical description of the physical process. This has led to the following limitations:

1. Due to the difficulty of writing a computer code to solve fluid dynamics calculations, one is usually limited to the capabilities of the commercially available codes.
2. The turbulence models currently in use do not accurately describe all possible situations.
3. The calculation of interphase boundary conditions between two fluids cannot always be handled correctly by the currently available codes.
4. In numerical methods the accuracy of a solution is dependent to an extent on the number of points in space and time for which calculations are made. Sometimes the number of points needed to achieve the desired accuracy in the calculations makes the cost of these calculations prohibitive. This is especially true in flow through very complex geometries.

References

(1) Anderson, D. A.; Tannehill, J. C.; Pletcher, R. H., Computational Fluid Mechanics and Heat Transfer; Hemisphere Publishing Corporation: New York, 1984.

(2) Bird, R. B.; Stewart, W. E.; Lightfoot, E. N., Transport Phenomena; John Wiley & Sons, Inc.: New York, 1960.

(3) Orr, C., Filtration, Principles and Practices; Parts I and II, Marcel Dekker, Inc.: New York, 1977.

(4) Patankar, S. V., Numerical Heat Transfer and Fluid Flow; Hemisphere Publishing Corporation: New York, 1980.

DYNAMIC AND STEADY STATE MATERIAL PROCESS MODELING

Use

Mathematical modeling and computer simulation have been widely used in the design, development and improvement of engineered material processes. The objective is to develop a mathematical model which will realistically describe the interactions among the process variables. Through experiments and the analysis of results, the model and its parameters may constantly be refined until it adequately predicts the performance of a process. This approach not only simplifies and speeds up design calculations, but generally provides more confidence and a better understanding of the process.

Typically, dynamic and steady state simulation are used in the following areas:

- Material purification and separation (single or multistage)
- Chemical reaction of materials

- Heat transfer.
- Momentum transfer.
- Mass transfer.
- Energy conversion.
- Materials handling.
- Process control.

Principles

The fundamental theories which underlie unit operation or process design are:

1. The first law of thermodynamics
2. The second law of thermodynamics
3. Newton's laws of motion.

The first law of thermodynamics yields equations for the conservation of mass and energy and can be combined with the laws of motion to yield equations for the conservation of momentum. These conservation equations often involve physical properties such as heat capacity, thermal conductivity, density, diffusivity, and viscosity. These properties may be divided into thermodynamic, transport and chemical-rate properties. The second law of thermodynamics yields equilibrium expressions and relationships between the thermodynamic, transport and chemical-rate properties. The equilibrium relationships indicate limitations of the physical and chemical changes which can be accomplished.

The type of equations (algebraic, differential, finite difference, etc.) which describe the system depend on both the process under investigation, and the details of its model. Most models for industrial processes are nonlinear and the solution can only be obtained via the use of applied numerical techniques with a computer.

Applications

When the appropriate physical properties and mathematical models are readily available, a unit operation or chemical process can be simulated on a computer.

The following examples represent some of the current applications:

- Prediction of equilibrium composition and temperatures for fractionators, absorbers, strippers, or flashes.
- Prediction of the temperature or composition profiles as a function of time/length in a chemical reactor.
- Estimation of the product temperatures and energy requirements in heat exchangers and furnaces.

- Prediction of power requirements, pressure drops, and staging/intercooling for compressors/pumps/blowers.
- Prediction of holdup time/sizing for storage vessels.
- Estimation of solid/liquid or solid/gas separation in hydrocyclones, cyclones, filters, crushers, or precipitators.
- Simulation of the response of a process subject to a control strategy.
- Equipment sizing and cost estimation.

Limitations

- The mathematical model may be unable to describe the real physical process. This can be due to the selection of an inappropriate model for the system or the model has an inordinate number of variables which requires an unreasonable processing time.
- Inaccuracies in the physical/thermodynamic properties can cause a realistic model to misrepresent the physical process.

References

(1) Bird, R. B.; Steward, W. E.; Lightfoot, E. N., Transport Phenomena; John Wiley & Sons, Inc.: New York, 1960.

(2) Danner, R. P.; Daubert, T. E., Manual for Predicting Chemical Process Design Data; A.I.C.H.E.: New York, 1983.

(3) Franks, R. G. E.; Modeling and Simulation in Chemical Engineering; Wiley-Interscience: New York, 1972.

(4) Hougen, O. A.; Watson, K. M.; Ragatz, R. A., Chemical Process Principles; John Wiley & Sons, Inc.: New York, 1959.

MOLECULAR MODELING

Use

Molecular Modeling provides a powerful new tool to the chemist for visualizing small, medium and large molecules individually or in combination. High resolution graphics and fast computers allow the user to build molecules in minimum energy conformations and view them from different directions in real time as ball and stick (Figure 1) or space filling (CPK, Figure 2) representations. Molecular dynamics can be studied by calculating the energy of a molecule or group of molecules as structural changes are imposed and representing the results visually (Figures 3 and 4). Molecular modeling has found particular use in the study of biomolecules by allowing visual representation of structure and reactivity relationships. Docking

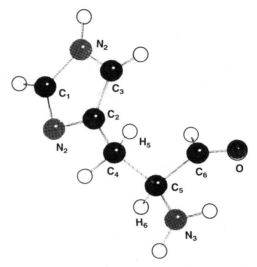

Figure 1. Ball and stick representation of histidine.

studies can be done where the structure and energy changes associated with enzyme/substrate interactions are calculated. The techniques developed for biopolymers are being extended to the study of organic polymers and inorganic materials such as zeolites.

Figure 2. Space filling (CPK) representation of tetraiodothyronine.

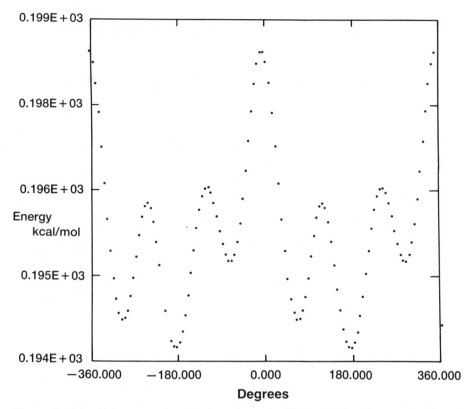

Figure 3. Potential energy plot rotation about the H_5-C_4-C_5-H_6 carbon-carbon bond of histidine (Figure 1).

Principle

Graphic representation of molecules gives the chemist a new tool to aid him in studying structure and reactivity. The molecular model can be constructed from crystallographic coordinates available from several data bases or by direct intervention by the chemist. In the latter mode the chemist builds a model in an interactive mode from atoms or molecular fragments available from fragment libraries. Molecular mechanics or quantum mechanics programs are then used to arrive at the best structure. The chemist is able to view the molecule from any direction by interactive rotation of the model about its coordinates. Molecules can be displayed as ball and stick or space filling models, or with surfaces showing charge distributions. Several molecules can be viewed at the same time and the system energy can be calculated as the molecules interact with one another. These molecular dynamics studies provide insight into molecular interactions in ground and transition states which determine structure and reactivity. As faster com-

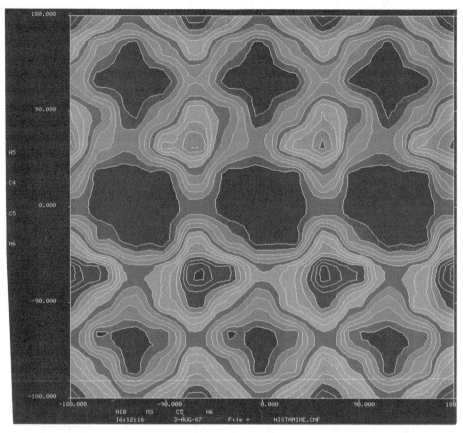

Figure 4. Contour potential energy plot for simultaneous rotation about the side chain C-C bond (y-axis) and C-N bond (x-axis) of histamine.

puters and more sophisticated modeling programs develop, it should be possible to design new materials with special properties for specific applications.

Applications

Molecular modeling has had a major impact on the design of new drugs and on the understanding of protein and enzyme structures and functions. It provides the researcher with the tools to visually display and manipulate the large numbers of atoms in a protein or enzyme and provides insight into structure reactivity relationships. Modeling has been used to study the effect of amino acid sequences on protein conformations, to investigate the interactions of proteins and enzymes with substrates, to determine the binding between donor and receptor molecules and to determine the charge distribution on an enzyme surface. Dynamic calculations allow the study of chain folding. Structural constraints derived from NMR and x-ray data can be included as parameters. The importance of water in determining biomolecule structure is also being investigated.

The techniques developed for modeling biopolymers are also being applied to the study of organic polymers. The conformations of repeating polymer chains and the intermolecular interactions between chains can be modeled. The effect of temperature on structure can also be simulated and the results compared to structural data derived from x-ray, IR and NMR spectroscopy. Information on polymer compatability and on the effect of additives such as plasticizers should also be attainable by molecular modeling.

Molecular modeling is easily applied to the viewing of any material where crystallographic coordinates are available. It should find wide application in the study of the structures of inorganic materials such as minerals, semiconductors and ceramics. Modeling of solid surfaces and their interaction with other solids and liquids should lend new insight into surface phenomena.

Limitations

Molecular modeling for display of complex molecules is limited by the resolution of the graphics display terminal and the size and speed of the computers used to generate and manipulate the graphics data. It is further limited by the quality of the energy minimization programs currently available. These programs are compute intensive and can use up large quantities of CPU time.

References

(1) Beveridge, D. L.; Jorgensen, N. L. ed., Computer Simulation of Chemical and Biochemical Systems; Ann. N.Y. Acad. Sci.: New York, Vol 482, 1986.

(2) Venkataraghavan, B.; Feldman, R. J. ed., Macromolecular Structure and Specificity: Computer Assisted Modeling and Applications; Ann. N.Y. Acad. Sci.: New York, Vol 438, 1985.

Acknowledgment

Figures 1–4 are by ChemGraf, created by E. K. Davies, Chemical Crystallography Laboratory, Oxford University and reprinted with permission of Chemical Design Ltd., Oxford.

Conversion Factors

Listed below are some common chemical, physical and metric units that are not acceptable in the SI Metric System, the acceptable SI equivalent,, and the appropriate conversion factor. A more complete list may be found in American National Standard Practice, ANSI Document No. Z210.1-1976, American National Standards Institute, New York, 1976.

To Convert From	To (SI Unit)	Multiply By[1]
AREA		
ft^2	square meter (m^2)	9.290 304 E-02
in^2	square meter (m^2)	6.451 600 E-04
ELECTRICITY AND MAGNETISM		
gauss[2]	tesla (T)	1.000 000 E-04
mho	siemens (S)	1.000 000 E+00
oersted	ampere per meter (A/m)	7.957 747 E+01
ohm centimeter	ohm meter ($\Omega \cdot m$)	1.000 000 E-02
ENERGY (INCLUDING HEAT, WORK AND POWER)		
Btu (thermochemical)	joule (J)	1.054 350 E+03
Btu/hr	watt (W)	2.928 751 E-01
Btu/$ft^2 \cdot h$	watt per square meter (W/m^2)	3.152 481 E+00
Btu/lb	kilojoule per kilogram (kJ/kg)	2.324 444 E+00
Btu/lb·°F	kilojoule per kilogram kelvin (kJ/kg·K)	4.184 000 E+00
calorie (thermo)	joule (J)	4.184 000 E+00
cal/g·°C	kilojoule per kilogram kelvin (kJ/kg·K)	4.184 000 E+00
cal/g	kilojoule per kilogram (kJ/kg)	4.184 000 E+00
cal (thermochemical)/min	watt (W)	6.973 333 E-02
electron volt	attojoule (aJ)	1.602 19 E-01
erg	joule (J)	1.000 000 E-07
erg/sec	watt (W)	1.000 000 E-07
HP (550 ft·lbf/s)	watt (W)	7.456 999 E+02
kilocalorie (thermo)	kilojoule (kJ)	4.184 000 E+00
kW·h	joule (J)	3.600 000 E+06
therm	joule (J)	1.055 056 E+08
FORCE		
dyne	newton (N)	1.000 000 E-05
kgf	newton (N)	9.806 650 E+00
lbf	newton (N)	4.448 222 E+00
FORCE PER UNIT LENGTH (SURFACE TENSION)		
dyne/cm	newton per meter (N/m)	1.000 000 E-03
LENGTH		
angstrom	nanometer (nm)	1.000 000 E-01
foot	meter (m)	3.048 000 E-01
inch	meter (m)	2.540 000 E-02
micron	micrometer (μm)	1.000 000 E+00
mil	millimeter (mm)	2.540 000 E-02
millimicron	nanometer (nm)	1.000 000 E+00
MASS		
ounce (avdp)	kilogram (kg)	2.834 952 E-02
pound (avdp)	kilogram (kg)	4.535 924 E-01
ton (short)	kilogram (kg)	9.071 847 E+02

To Convert From	To (SI Unit)	Multiply By[1]
MASS PER UNIT VOLUME		
g/cm^3	kilogram per cubic meter (kg/m^3)	1.000 000 E + 03
lb/ft^3	kilogram per cubic meter (kg/m^3)	1.601 846 E + 01
lb/US gal	kilogram per cubic meter (kg/m^3)	1.198 264 E + 02
MASS PER UNIT TIME (FLOW RATE)		
lb/h	kilogram per second (kg/s)	1.259 979 E−04
ton (short)/h	kilogram per second (kg/s)	2.519 958 E−01
MOLAR MASS		
lb/mol	kilogram per mol (kg/mol)	4.535 924 E−01
PRESSURE AND STRESS		
atmosphere (stand)	kilopascal (kPa)	1.013 250 E + 02
bar	pascal (Pa)	1.000 000 E + 05
dyne/cm^2	pascal (Pa)	1.000 000 E − 01
inch of mercury (32°F)	kilopascal (kPa)	3.386 38 E + 00
inch of water (60°F)	kilopascal (kPa)	2.488 4 E − 01
millimeter of Hg(0°C)	pascal (Pa)	1.333 22 E + 02
psi (lbf/in^2)	kilopascal (kPa)	6.894 757 E + 00
torr (mm Hg, O°C)	pascal (Pa)	1.333 22 E + 02
TEMPERATURE		
degree Celsius	kelvin (K)	$t_K = t_C + 273.15$
degree Fahrenheit	degree Celsius (°C)	$t_{°C} = (t_{°F}-32)/1.8$
TORQUE		
dyne•cm	newton-meter (N•m)	1.000 000 E−07
lbf•ft	newton-meter (N•m)	1.355 818 E + 00
VELOCITY		
ft/min	meter per second (m/s)	5.080 000 E−03
ft/s	meter per second (m/s)	3.048 000 E−01
VISCOSITY		
centipoise	pascal second (Pa•s)	1.000 000 E−03
centistokes	square meter per second (m^2/s)	1.000 000 E−06
VOLUME		
ft^3	cubic meter (m^3)	2.831 685 E−02
gallon (US liquid)	cubic meter (m^3)	3.785 412 E−03
pint (US liquid)	cubic meter (m^3)	4.731 765 E−04
quart (US liquid)	cubic meter (m^3)	9.463 529 E−04
VOLUME PER UNIT TIME (FLOW RATE)		
gallon (US liquid)/min	milliliter per second (mL/s)	6.309020 E + 01
PHYSICAL CONSTANTS		
Avogadro's number	6.022 17 E + 23 molecules/mol	
Boltzmann's constant	1.380 62 E − 23 J/K	
Faraday's constant	9.648 67 E + 04 C/mol	
Gas constant	8.314 34 E + 00 J/K•mol	
Gas constant	8.314 34 E + 00 Pa•m^3/K•mol	
Planck's constant	6.626 20 E − 34 J•s	

Notes:

(1) The conversion factor is followed by the letter E (for exponent), a plus or minus symbol, and two digits which indicate the power of 10 by which the number must by multiplied to obtain the correct value. For example, 9.290 304 E − 02 is 9.290 304 x 10^{-2}.

(2) The relationship between flux density B and field strength H in SI units is defined to be $B = \mu_o H + J$ where $\mu_o = 4\pi \times 10^{-7}$ H/m(permeability constant) and J = magnetic polarization (expressed in teslas).

Index

Absorption coefficients 17, 127
Absorption edge 127, 128
Accurate mass measurement 47, 48
Algorithms 291
Amorphous solids 118
Analysis of degradation products 14, 19, 46
Analysis of ionic species 46, 76, 91, 100
Angular dependent XPS (ADXPS) 197
Anisotropy 217, 225
Angular divergence 134
Applied finite element analysis 291
ASTM 274
Atomic absorption spectroscopy 4, 101
Attenuated total reflection (ATR) 14, 16, 17
Auger electron spectroscopy (AES) 4, 168, 177
Automated C, H, N, and O analysis 89
Automatic image analysis 150, 160

Biaxial elongational viscosity 229
Binding energy 167
Biological sections 138
Biopolymers 29, 54
Birefringence 138
Bragg condition 115
Branching 14, 29, 63, 83, 238
Bremsstrahlung isochromat spectroscopy (BIS) 193
Bright field imaging 151, 152

Capillary flow 229, 239
Characteristic emission lines 131
Characteristic x-rays 145
Chemical analysis 85
Chemical ionization (CI) 45
Chemical shifts 29, 128
Chromatographic separation 16, 51
Chromatography 50, 61
Chromatography-mass spectrometry 50
Classical chemical analysis 85
Coefficient of thermal expansion 269, 270, 271
Cohesive energy density 264
Colligative properties 235, 244
Collision induced dissociation 57
Combustion analysis 89
Compatibility 264
Compliance 287
Composites 143, 217, 222, 225
Compositional determination 131, 132
Compression 274
Computational fluid dynamics modeling 294
Computer 291
Computer-based model 293, 294, 297, 299
Conformational analysis, 14, 20, 21
Conformational isomers 20
Contact angles 262, 267
Continuum mechanics 273
Copolymers 14, 17, 29, 31, 82, 83, 139, 140, 154, 157
Corrosion 55, 144, 145
Coulter principle 251, 252
Crack propagation 281, 285
Crosslink density 243
Cryosectioning 150, 154, 165
Crystallinity 117, 119, 271
Crystalline index 14, 120
Crystalline modulus 119
Crystallite size 117, 119
Crystallization 212, 269
Cutting/polishing techniques 150, 164, 165
Cyclic voltammetry 96
Cycling stress-strain diagram 279

Dark field 137, 139, 151, 152, 156
Density 261, 270

Density of occupied and unoccupied states 128, 195
Depth profiling 55, 180, 181, 183, 189, 190
Detection limits by ion-selective electrodes 91
Diamond wheel cutting 164
Dielectric thermal analysis (DETA) 227
Differential pulse polarography (DPP) 96, 98
Differential scanning calorimetry (DSC) 3, 211, 269
Differential thermal analysis (DTA) 3, 205, 211, 269
Diffraction 115, 119, 150, 151
Diffusant gas or liquid 288
Diffusion coefficient 256, 289
Digital imaging 151, 160
Dilatometry 217, 219, 269, 270
Dimensional changes 217, 269
Direct-reading emission spectrograph 105, 106
Dispersion 138
Dropping mercury electrode 96
Ductility 273
Dynamic and steady state modeling 297
Dynamic mechanical analysis (DMA) 223
Dynamic modulus 223, 224, 233, 282, 284
Dynamic response 284
2D-NMR (Nuclear Magnetic Resonance) 36
2D-NOE (Nuclear Overhauser Effect) 37

Elastically backscattered electrons 142
Elasticity 231, 234
Electroactive functional groups 98
Electrochemical reaction 96, 165
Electrolyte 96, 165
Electrolytic conductivity detectors 62
Electron beam 142, 145, 150, 160
Electron capture 63
Electron diffraction patterns 151, 152, 154
Electron impact (EI) 45
Electron microscopy 142, 150
Electron paramagnetic resonance spectroscopy (EPR) 40
Electron probe microanalysis (EPMA) 142, 145
Electron spectroscopy for chemical analysis (ESCA) 167
Electropolishing 150, 165, 166
Elemental composition 3, 51, 85, 89, 90, 131, 145, 151, 152
Emission spectrographic analysis 104
End group analysis 14, 29, 30, 88
Energy dispersive x-ray fluorescence (EDXF) 131, 145, 146, 152

Energy level diagram 177
Energy release rate 285
Epidiascope 160
Exact mass measurement 47, 58
Extended x-ray absorption fine structure spectroscopy (EXAFS) 127

Failure analysis (delamination) 139, 144
Fast atom bombardment (FAB) 53
Fatigue analysis 278
Federal specifications 274
Fiber denier 161
Fibers 119, 120, 121, 123, 138, 150, 152, 155, 217, 224, 225, 226, 233, 273
Fick's second law 289
Field flow fractionation 245
Finite element modeling 291
Flame emission spectrometry 107
Flame ionization detectors 67
Flammability 206
Flexing 274
Fluid dynamics modeling 294
Fourier transform 15, 27, 68, 128
Fourier transform infrared spectroscopy 15
Fracture toughness 285
Fragment ions 46, 57, 58
Frequency distribution determination 161
Functional group analysis 87
Fundamental parameters software 132

Gallium ion source 186
Gas adsorption 258
Gas chromatography (GC) 61
Gas permeability 288
Gas phase collision 46, 57, 58
Gaseous hydride method 102
Gel permeation chromatography 81, 251
Glass transition 5, 14, 19, 137, 212, 270
Gradient elution 67
Graphite furnace 77, 102
Gravimetry 86
Grey level 160

High energy electrons 145, 150
High molecular weight fraction 234
High performance liquid chromatography (HPLC) 2, 66

Hydrolysates 63
Hydroxyl groups 88

Impact strength 284
Impurities in drinking water 97
Inclusion identification 16, 21, 144, 145
Index of crystallinity 14, 120
Inductively coupled argon plasma emission spectroscopy (ICAP) 4, 101, 109
Inelastically scattered electrons 142, 197
Infrared spectroscopy 2, 13
Inorganic structure analysis 3
Inorganic lattices 14
Instrumented impact analysis 282
Interplanar spacings 116
Intrinsic viscosity 239
Ion beam 46, 57, 164, 165, 186, 189
Ion chromatography 76
Ion exclusion chromatography 78
Ion induced secondary electron images 185
Ion milling 150, 165
Ion pair reverse-phase high pressure liquid chromatography 68
Ion selective electrode analysis 3, 4, 91
Ionic compounds 55
Ionic species 76, 77, 91, 99, 100
Isotachophoresis analysis 99
Isothermal TGA 205

J-integral 286

Kinetic studies 14, 206

Lamellar structure 124, 153, 157
Laminated films 139
Lateral elemental distribution 142, 145, 182, 188
Lattice constants 118
Lattice spacing 151
Lattice vibrations 20
Light scattering 241, 242, 251, 255
Light transmission 24, 137
Line profiling 145, 177
Linear elastic fracture mechanics (LEFM) 285
Liquid chromatography (LC) 66
Liquid metal field ion emission probe 186

Liquid (mobile) phase 66
Liquid permeability 288
Liquid scintillation counting 94
Lower limit of detection (LLD) 102, 106, 108, 110, 132
Low and high temperature diffractometry 118
Low resolution mass spectra 45, 47
Low voltage imaging 143, 144

Macrofractography 280
Magic angle spinning 33
Magnetic domains 144
Magnetic moment 27, 40
Magnetic transitions 41, 206, 207, 212
Mapping 145, 146, 177, 182, 183
Mark-Houwink equations 239
Mass spectrometry 45
Material process modeling 297
Mean free path 197, 198
Mechanical polishing 139, 164
Mechanical properties 233, 273
Mechanical properties of solids and fluids 233
Mechanical spectrometry 223
Mechanical testing 273
Melting 213, 216, 271
Melt elasticity 232, 234
Melt heterogeneities 232
Membrane osmometer 236
Mercury intrusion porosimetry 261
Metallic alloys 102, 106, 110, 143, 144, 145, 207, 208, 212, 214, 222, 223, 224, 225
Metallographic investigations 139, 144, 145
Microcrystalline 118, 151, 152
Microdiffraction 134, 151, 152
Microfractography 280
Microscopy 137
Microscopy specimen preparation techniques 150, 163
Microstructures 14, 137–163, 273, 274
Modeling 6, 119, 293–303
Modification ratio 161, 163
Modulus 6, 119, 223, 224, 233, 273, 274, 282, 284
Molecular entanglements 231
Molecular ion 46
Molecular modeling 119, 299
Molecular orbitals 193

Molecular spectroscopy 13
Molecular vibrations 13
Molecular weight 4, 5, 14, 30, 45, 46, 47, 49, 52–57, 81, 229–249
Molecular weight distribution 5, 54, 81, 232, 234, 248, 257
Molecular weight between crosslinks 244
Molecular weight of polymers 14, 229–249
Monomers 63, 236
Morphology 133, 150, 152, 219

Near-infrared (diffuse) reflectance analysis (NIRA) 24, 25
Near infrared spectroscopy 24
Neutron activation analysis 4, 112
Newtonian fluid 229–232
Non-Newtonian behavior 231
Nonvolatile mixtures 51, 53
Nuclear magnetic resonance spectroscopy (NMR) 2, 27–40
Nuclear magnetic resonance-solutions 27
Nuclear magnetic resonance-solids 33
Number average molecular weight 14, 30, 235, 236
Numerical methods 291, 294, 295, 298

Occupied and unoccupied density of states 193
Olefinic unsaturation 48
Oligomeric compounds 5, 54, 83
Oligopeptides 38, 54
Oligonucleotides 38, 54
Optical emission spectroscopy 4
Optical fiber couplers 144
Optical gradient techniques 252
Optical microscopy 4, 137, 163
Order-disorder transitions 118
Organic structure analysis 2
Orientation 14, 119, 138, 217
Oxidation state 127
Oxidative combustions 89
Oxidative stability 206
Oxygen transmission rate (OTR) 289

Paper chromatography (PC) 67
Parr oxygen bomb 90
Particle agglomerations 143
Particle size analysis 151, 161, 242, 251–258
Particle size distribution 83, 151, 161, 251–258
Particulate materials 138, 150, 163

Partition chromatography 67
Permeability 251, 258, 288
Petrographic thin sections 138
Phase composition 119, 135, 143, 151, 152
Phase contrast illumination 137, 139
Phase identification 118, 139, 143, 151, 152
Phase inversion 152, 155
Phase transitions 118, 139
Phenolics 54, 100
Photoacoustic spectroscopy (PAS) 16
Photodiode-array technology 67
Photoelectrons 167
Photoelectron "take-off" angle 197
Photoionization 62
Photometry 86
Photon correlation spectroscopy 251, 255
Physical properties 251
Plasma discharge ionization 45
Polarography 96
Polarized light 14, 138
Polycrystalline materials 115, 139, 142, 160
Polydispersity 235, 241
Polymer additives 54, 83, 124, 206, 237, 279
Polymer blends 124, 137, 150
Polymer characterization 4
Polymer films 13, 137
Polymer melts 234
Polymer stabilizers 54, 206
Polymerization 41, 206, 212
Polymerization rates 212, 232
Polymers 4, 14, 19, 27, 40, 61, 119, 137, 150, 160, 205–245, 264–271
Polymers in solution 27, 124, 235, 238, 241, 243, 264
Pore size 258, 261
Pore size distributions 258, 261
Power law 231
Preparative liquid chromatography 66
Priority pollutants 51
Problem solving 7
Process control 232
Process modeling 297
Processing aids 54
Processing conditions 232
Production of secondary electrons 142, 177, 178
Proteins 38, 54

Pseudoplastic behavior 230
Purity determination 87, 206
Pyrolysis 63, 48

Quadrupole mass filters 57

Radial distribution function 128
Radioactive tracer analysis 94
Radius of gyration 242
Raman spectroscopy 2, 19
Rates of crystallization 212
Reaction kinetics 14, 28, 29, 206, 256
Reaction kinetics modeling 297
Refractive-index 67, 82, 252
Real-time computer data acquisition and control 45, 299
Residual stress 135
Resistance to perforation 284
Resonance Raman effect 21
Rheology of fluids 229
Rheometers 229
Rigidity of materials 273
Rocking curves 135

Sampling depth 16, 55, 197
Scale effect 281
Scanning auger microscopy (SAM) 177
Scanning electron microscopy (SEM) 4, 142, 143
Scanning transmission electron microscopy (STEM) 150, 151, 160
Scanning tunneling microscopy 200
Schöniger oxygen flask combustion 89
Scientific computation 6, 291
Second virial coefficient 235, 236, 242
Secondary electron image 142, 185
Secondary emission 142
Secondary ion mass spectrometry (SIMS) 4, 185
Sectioning 150, 164
Sector magnet 57
SEM or STEM images 144–162
Semiconductors 152, 169, 182, 188
Semiconductor interfaces 152
Semicrystalline polymers 4, 124
Separation of overlapping peaks 117
Sequence distribution in copolymers 32
Shear 274

Shear rates 234
Simulation 292, 293, 303
Size of globular particles 124, 125
Skeletal density 261
Short bar chevron notch (SBCN) 286
Short rod/short beam 286
Sintering 219
Size distribution analysis 124, 153, 161, 251–263
Small angle x-ray and neutron scattering 124
"Soft" ionization 45, 53, 57
Software 295
Solid solutions 118
Solid state NMR 33
Solubility 264
Solubility parameter 264
Sonic modulus 223
Spatial resolution 137, 141, 142, 150, 179
Specific viscosity 239
Specificity and sensitivity of detection of radioactive isotopes 94
Spherulitic development 138
Spin decoupling 29
Spin-spin coupling 28, 29
Spot analysis 145, 151, 177
Spraying 150
Sputtering 164
Square wave voltammetry 96
Staining techniques 150, 152, 154
Steric purity 264
Stiffness 6, 225
Storage modulus 224, 234
Strength of materials 262, 274
Stress intensity factor 285
Stresses and strains 7, 274
Stripping voltammetry 96
Structural analysis 38, 301
Sub-sieve sizer 252, 253
Sulfonates 54, 78
Superconductors 3
Supercritical fluid chromatography 71
Surface adsorbed layers 54, 258
Surface adsorbed species 54, 260
Surface analysis 167
Surface composition 14, 55, 167
Surface discontinuities 139

Surface effects 55
Surface energy 266
Surface tension 267
Surface topography 142
Surfactants 54
Synchrotron x-ray sources 133

Tacticity 29
Tandem mass spectrometry (MS-MS) 57
Tension 274
Tension-compression 233
Textile finishes 143
Theory of crack propagation 285
Thermal analysis 205
Thermal conductivity 222, 298
Thermal expansion coefficients 118, 217
Thermal stability 206, 232
Thermal stage 137
Thermodynamics modeling 298
Thermodynamic parameters 264
Thermogravimetric analysis (TGA) 205
Thermomechanical analysis (TMA) 217, 269
Thermospray mass spectrometry 51
Thin layer chromatography (TLC) 67
Three point bending 233
Torsion 233, 279
Total organic carbon 89
Trace analysis 3, 49, 61, 62, 63, 69, 77, 92, 96, 102, 106, 108, 111, 112, 131
Transition temperature 137, 211
Transmission electron microscopy 4, 150, 165
Transmission rate 288
Transmitted electron 150
Transmitted and reflected light 138, 139
Transport properties 298
Turbidimetry 86
Two-dimensional position sensitive detector 119
T1 (Spin-lattice relaxation time) 29
T2 (Spin-spin relaxation time) 29

Ultraviolet absorption 24
Ultraviolet spectroscopy 24, 193
Unimolecular decompositions of ions 46
Unoccupied electronic states 127, 193

Vacuum evaporation 164
Valence band(s) 193
Vapor permeability measurements 236
Vapor pressure osmometer 236
Viscoelastic characteristics of liquids, solutions and melts 229
Viscoelastic properties 223
Viscosity 5, 234, 238, 298
Viscosity and molecular weight of polymers 229
Viscosity average molecular weight 238
Viscosity of polymer solutions 238
Visible light 24, 137
Visible spectroscopy 24
Voltammetry 96, 97
Volatile complex mixtures 62
Volumetry 86
Void content 124, 143, 161

Water molecules (D_2O) in nylon 6 125
Water vapor transmission rate 289
Wavelength dispersive spectrometers 131, 145
Weight average molecular weight 241
Wilhelmy balance 267
Work function (θ) 195

X-ray analysis 115
X-ray absorption near edge spectroscopy (XANES) 127
X-ray diffraction-polymers 3, 119
X-ray distribution mapping 145
X-ray fluorescence (XRF) 3, 131
X-ray photoelectron spectroscopy (XPS) 4, 6, 167
X-ray powder diffraction 115
X-ray spectra 145
X-ray topography 135